浙江省建筑设计研究院作品选

ZIAD

SELECTED WORKS

1952-2006

中国建筑工业出版社
CHINA ARCHITECTURE & BUILDING PRESS

图书在版编目(CIP)数据

浙江省建筑设计研究院作品选／《浙江省建筑设计研究院作品选》编委会编. –北京：中国建筑工业出版社，2007
ISBN 978-7-112-09437-0

Ⅰ. 浙… Ⅱ. 浙… Ⅲ. 建筑设计–作品集–中国–现代
Ⅳ. TU206

中国版本图书馆CIP数据核字(2007)第089781号

责任编辑：郭洪兰
责任校对：兰曼利　王金珠

浙江省建筑设计研究院作品选 1952–2006
ZIAD
*
中国建筑工业出版社出版、发行（北京西郊百万庄）
各地新华书店、建筑书店经销
北京方舟正佳图文设计有限公司制版
北京盛通印刷股份有限公司印刷
*
开本：787 × 1092毫米　1/10
印张：29　字数：588千字
版次：2007年12月第一版
印次：2007年12月第一次印刷
印数：1-2000册
定价：280.00元
ISBN 978-7-112-09437-0
(16101)

回顾历史　展望未来

浙江省建筑设计研究院成立于1952年，初始为浙江省建工局技术设计处，1956年更名为“浙江省城市设计院”，虽然只有四、五十名设计人员的规模，但其时却是浙江省惟一的设计单位，为浙江省经济恢复时期的工程建设做出了很多贡献。1958年后，为适应经济建设的需要，更名为“浙江省工业设计院”，设计范围逐步扩大，涉及到城市规划、工业与民用建筑、水利、电力、化工、冶金、煤炭、工程勘测等工程勘察设计领域，职工人数达800余人，是当时省内规模最大、设计范围最广的设计院，为浙江省“一五”至“五五”计划的完成付出了艰辛的努力。1980年更名为以从事工业与民用建筑设计为主的“浙江省建筑设计院”。1992年再次更名为浙江省建筑设计研究院至今。

目前我院现有职工近500名。其中国家工程设计大师2名；教授级高级工程师、教授级建筑师35名；国家一级注册建筑师、一级注册结构工程师112名；高级工程师、高级建筑师100余名；工程师、建筑师近200名。自1952年建院以来，共承担国内外各类工业与民用建筑项目设计一万余项，有300余项分别获国家、部、省级优秀设计奖，80余项科研项目分别获国家、部、省级科技成果奖、科技进步奖、全国计算机优秀软件奖和国家专利等。1992年首批入选中国勘察设计单位综合实力百强之一，1996年被国家建设部授予“全国建筑节能先进集体”，1997年被国家建设部授予“精神文明建设先进单位”，2000年被国家科学技术部授予“全国CAD应用工程示范企业”，2006年获建设部颁发的“‘十五’全国建设科技进步先进集体”。

五十多年来，我院经历了组建、发展、成熟的三个阶段，体会到创业的艰辛，也共享丰收的喜悦。多年来虽数易院名，但全院职工始终不渝地坚持“设计创新，质量创优，诚信求实，团结敬业”的精神，在各个历史年代创作出了一大批符合时代气息、特色鲜明的代表性建筑。

早在20世纪50～70年代，由我院设计的杭州饭店（现名杭州香格里拉饭店）、省府一号、二号楼、杭州市人民大会堂、浙江体育馆（现名杭州体育馆）、杭州笕桥民航候机楼、杭州剧院、杭州西湖国宾馆、西子宾馆、浙江宾馆、望湖宾馆等一大批符合“适用、经济、美观”和时代感、标志性都很强的大型民用公共建筑陆续建成。同时还设计和建成了反映浙江省工业建设从无到有、从小到大的杭州钢铁厂、杭州重机厂、浙江麻纺厂、杭州第一、第二棉纺厂、诸暨绢纺厂、杭州都锦生丝织厂、嘉兴毛纺厂、诸暨毛纺厂、余姚化纤厂等一大批大、中型工业骨干工程。承担了冶金、机械、电子、化工、轻工、纺织、煤炭、医药、食品、军工等行业大量的工业建筑以及遍及全省各地的大、中型肉类、水产、蛋品、水果等冷藏库工程。

特别是实施改革开放以来，国家经济建设快速发展，我院也迎来了难得的历史发展机遇，设计了浙江世贸中心、西藏体育馆、浙江图书馆、杭州游泳馆、上海国际会议中心、上海金钟广场、杭州黄龙体育中心（主体育场、主体育馆）、上海外贸学院、上海东方绿舟、杭州萧山国际机场、山东大厦、西湖凯悦国际大酒店、上海视觉艺术学院、中国恐龙馆、西湖文化广场等一批在国内外产生影响的建筑精品，并因此分别获得上海市50年经典建筑中的金奖、铜奖，国家70年代优秀设计银奖和省、部级优秀设计奖。

我院的科研项目也是硕果累累。预应力混凝土空心桩、预制钢筋混凝土槽板组合H型桩应用研究、球团烧结竖炉设计研究、低露点空气干燥机组、弹性地基上基础的合理设计及桩土共同作用分析研究、深基坑支护设计计算研究、建筑工程应用地下连续墙的设计计算理论与施工技术的研究等项目均通过鉴定并分别获得浙江省科学技术进步奖、省建设科学进步一、二、三等奖和冶金工业部科技成果一等奖等。此外还先后主持编制了浙江省标准《地基基础设计规范》、《基桩低应变动力检测技术规范》、《建筑基坑工程技术规范》、《混凝土多孔砖设计与施工技术规程》和国家行业标准《办公建筑设计规范》等。

回顾过去的55年，在省建设厅等上级部门的领导与支持下，我院历届领导班子团结奋斗，组织和依靠全体设计人员，取得了十分喜人的成绩。特别是老一辈的建筑师、工程师们为我院留下了众多建筑精品和宝贵经验，也为我们留下了团结奋斗、精益求精、乐于奉献的精神财富。进入21世纪，面临着更多的机遇与挑战，我院将继续认真贯彻执行国家的建设方针和政策，努力落实科学发展观，不断学习、借鉴世界各国先进的建筑设计理念和继承发扬我国的优秀建筑文化，创作符合我国国情和地域文化的新世纪精品建筑，为我国经济建设、城乡建设、工程建设做出更多贡献。

2007年3月16日

Reviewing the Past and Looking Forward to the Future

Established in 1952, Zhejiang Province Institute of Architectural Design & Research (ZIAD) was firstly named Technical Design Department of Zhejiang Province Building & Engineering Administrative Bureau and renamed to Zhejiang Province Urban Design Institute in 1956 which holds only forty to fifty designers. But it was the only design unit in Zhejiang Province at that time which made great contributions to the construction of Zhejiang during its economic recovery period. Following the economic construction since 1958, it was changed its name to Zhejiang Province Industrial Design Institute and expanded its service scope to urban planning, geotechnological survey and design for industrial and civil building, water conservation, electric power, chemical engineering, metallurgy, coal industry and engineering exploration, etc. It consisted of over 800 staff and was renowned by its largest scale and wide-ranging disciplines in Zhejiang Province. ZIAD has worked hard on the execution of First Five-Year Plan and Fifth Five-Year Plan of Zhejiang Province. In 1980, it was renamed to Zhejiang Province Architectural Design Institute mainly involving in the design of industrial and civil building. In 1992, it was once more renamed to Zhejiang Province Institute of Architectural Design & Research and never changed it up to now.

ZIAD has approximate 500 staff including two national design masters, 35 professorate senior architects and engineers, 112 first-class state registered architects and engineers, over 100 senior architects and engineers and more than 200 architects and engineers. Since its founding in 1952, ZIAD has been involved in the design of over 10 000 civil and industrial projects at home and abroad. Over 300 projects have been awarded the National, Ministerial or Provincial Excellent Design Prize and more than 80 S&T programs have received the National, Ministerial or Provincial Scientific & Technological Achievement Prize, Scientific & Technological Advancement Prize, State Advanced Computer Software Prize and national patent. In 1992 ZIAD was chosen as one of the first to be honored Top 100 Survey and Design Institutes in China. In 1996, ZIAD was given the title of State Advanced Collective in Promoting Building Energy Conservation and granted the honor of Advanced Enterprise in Constructing Spiritual Civilization in 1997 by the Ministry of Construction. In 2000 it was honored National CAD Application Demonstration Enterprise by the State Ministry of Science and Technology. In 2006 it was honored Tenth Five-Year-Plan National Construction Scientific & Technological Advanced Collective.

During the past 50 years, ZIAD experienced all kinds of hardships in starting its business and the joy when bumper harvest was coming. It went through three stages from organization, development to maturity. Although it was renamed many times, all staff has been keeping the spirit and principle of Design Innovation, Quality Improvement, Upholding Integrity & Truth and Promoting Solidarity and Work Respect. During every historical period, ZIAD creates a number of representative buildings that convey sense of times and show strong identity.

In the early 1950'to 1970', ZIAD has been involved in the designs for large numbers of civil and public buildings which correspond to the principle of Utility, Economy and Beauty, convey sense of times and exhibit themselves as signature buildings. They were Hangzhou Hotel (now Shangri-La Hotel Hangzhou), No.1 & No. 2 Building of Provincial Government, Hangzhou People's Convention Hall, Zhejiang Gymnasium (now Hangzhou Gymnasium), Hangzhou Jian Qiao Airport Terminal, Hangzhou Theater, Hangzhou Xihu State Guesthouse, Xizi Hotel, Zhejiang Hotel and Hangzhou Lakeview Hotel, etc. The other key industrial buildings which were designed by ZIAD and completed include Hangzhou Iron & Steel Plant, Hangzhou Heavy Machine Plant, Zhejiang Linen Textile Factory, No.1 & No.2 Hangzhou Cotton Mill, Zhuji Silk Spinning Factory, Hangzhou Dujinsheng Silk Weaving Mill, Jiaxing Woolen Mill, Zhuji Woolen Mill, Yuyao Fiber Mill. Moreover ZIAD designed large amounts of industrial buildings in the field of metallurgy, machinery, electronic industry, chemical industry, light industry, textile industry, coal industry, medicine, food and war industry, etc. It also designed cold store projects for the processing plant of meat, aquatic product, egg and fruit, etc.

With the reform and opening-up policy, China has witnessed its most rapid development. ZIAD meets the historical opportunities and creates a number of excellent designs including Zhejiang World Trade Center, Tibet Gymnasium, Zhejiang Library, Hangzhou Natatorium, Shanghai International Convention Center, Golden Bell Plaza Shanghai, Dragon Sports Center Hangzhou (Main Stadium & Gymnasium), Shanghai University of Foreign Trade, The Oriental Green Boat Shanghai, Hangzhou Xiaoshan International Airport, Shandong Building, The Hyatt Regency Hangzhou Hotel, Shanghai Institute of Visual Art, Chinese Dinosaur Museum Changzhou and Westlake Cultural Plaza Hangzhou, etc. Those impressive and quality architectures in the domestic and worldwide market were thus awarded the Golden and Copper Prize of Shanghai Classical Architecture in 50 Years, Silver Prize of National Excellent Design in 1970's and Provincial and Ministerial Excellent Design Prize.

In addition, ZIAD had great accomplishments in scientific research field including Prestress Concrete Hollow Pile, Study on the Application of Precast RC Channel Slab Composite H-Typed Pile, Study on Sintering & Pelletizing Shaft Furnace Design, Low-dew-point Air Dryer Unit, Reasonable Design for Foundation on Elastic Soil and Analysis on Pile-soil Interaction, Study on Design Calculation of Retaining of Deep Excavation, Study on Design Calculation Theory and Construction Method of Diaphragm Wall Applied to Building Engineering. All of them have passed the expert assessment and evaluation and have been awarded the Science and Technology Advancement Prize Zhejiang Province, First/Second/Third Prize of Construction Science Progress Zhejiang Province and First Prize of Science & Technology Achievement by the Ministry of Metallurgical Industry. Furthermore ZIAD prepared and compiled codes and standards applied in Zhejiang including the Code for Design of Soil Foundation, Technical Code of Low-Strain Dynamic Test to Pile Foundation, Technical Code of Building Foundation Pit, Regulation of Design of Concrete Perforated Brick and Construction Technology and Office Building Design Code which is served as state technical standard.

Reviewing the past 55 years, under the leading and support of Construction Bureau Zhejiang Province, all the Party and the administrative leaders of ZIAD are united in hard work, organize and depend all designers and staff to achieve gratifying successes. Especially those architects and engineers of old generation leave us not only the significant architectural buildings and valuable experiences but also the spirit wealth of solidarity and hard-working, constant perfection and dedication. Once entering the 21st century, we'll meet even more opportunities and challenges than before. ZIAD will keep on implementing the state policies to construction, strive to implement the scientific development viewpoint, keep on studying the worldwide advanced architectural design ideas, carry forward and develop the domestic excellent architectural cultures to provide high-quality works in new century which respect Chinese context and pay homage to the indigenous cultural heritage so as to make more contributions to the national economic construction, urban construction and building development.

2007.3.16

目　录

Contente

Medical Care Building

Residential Building

综合建筑
Multi-Functional Building

杭州西湖文化广场

Hangzhou Westlake Cultural Plaza, Hangzhou

西湖文化广场是目前浙江省最大的综合性文化类建筑。主体塔楼为俯瞰市区的地标建筑。总体的弧形布局与大运河弯流合围着位于城市中轴线上的主广场，$10hm^2$的室外环境用地中，由水岸、绿地、景观桥面、下沉庭院等多种室外场所组成了丰富的文化休闲环境。富有激情的外形也蕴含着丰富的内部空间，高技术与艺术的结合为市民创造了集科普、交流、展示、培训、影视演艺和旅游娱乐为一体的文化世界。

Prominently located in the north bank of Beijing-Hangzhou Grand Canal, Hangzhou Westlake Cultural Plaza is designed as the largest comprehensive cultural facility with its main tower structure dominating the view of the city. Its crescent-shaped plan complemented into the Grand Canal embraces the main plaza that is located on the main axis of the city. The 10-hectare site is composed of the water bank, green area, pedestrian bridge and sunken court,which act as a varied cultural and relaxing area. The exterior facades are sensitively sculptured to reflect its various interior layout. The combination between the new technology and the expressive form are proven to provide ideal cultural amenities accommodating the science, communication, exhibition, training, cinema, performance, tourism and entertainment facilities for the local citizen.

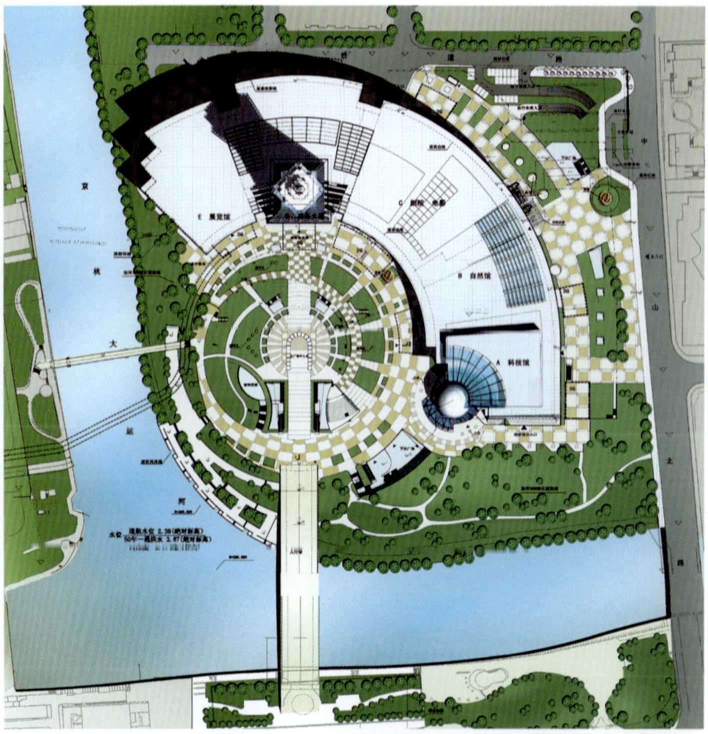

建 设 单 位：西湖文化广场建设指挥部
功 能 用 途：综合性文化广场
设计 / 竣工年份：2003 年 /2007 年
建 设 地 点：杭州市武林广场北
总 建 筑 面 积 ：373000m^2
总建筑高度 / 层数：170m/41 层
结 构 形 式：框剪结构

Client: Hangzhou Westlake Cultural Plaza Construction Headquarter
Function: Comprehensive Cultural Plaza
Designed/Completed: 2003/2007
Location: Wulin Square(North), Hangzhou
Total Floor Area: 373,000m^2
Total Height/Floor: 170m/ 41 Floors
Type of Structure: Framed Shear Wall Structure

上海国际会议中心

Shanghai International Convention Center,Shanghai

上海国际会议中心位于上海浦东小陆家嘴浦江东岸，占地4.5hm²，建筑面积11万m²，地上11层，主要功能为国际会议设施和五星级酒店。会议中心拥有从18人～800人的各类会议厅以及4000m²的大型多功能厅，五星级酒店拥有300间客房以及配套服务设施。建筑高度40m，大球体直径50m，小球体直径38m。全球财富论坛曾在此召开，APEC会议部分活动曾在此举行。

Located to the northwest of the Oriental Pearl Tower, the project occupies a total site area of 4.5 hectares and a total floor area of 110,000.square meters with 11 floors aboveground encompassing the international convention facilities and a 5-star hotel. The convention center is provided with meeting venues of various sizes to accommodate 18 to 800 guests and a ballroom of 4,000 sq.m. . The luxurious hotel is distributed by 300 guest rooms and other back-of-house premises. The total building is 40 meters high with two sphere structures, the larger one is 50 meters in diameter and the smaller one is 38 meters in diameter. Shanghai International Convention Center hosted the 1999 Shanghai Fortune Globe Forum and APEC China 2001.

上海国際會議中心

建 设 单 位：上海市广播电视局
功 能 用 途：会议、宾馆
设计 / 竣工年份：1996 年 /1999 年
建 设 地 点：上海浦东小陆家嘴
总 建 筑 面 积 ：110000m²
总建筑高度 / 层数：40m/11 层
结 构 形 式：框架结构、钢结构

Client: Shanghai Broadcasting and TV Bureau
Function: Convention Hall, Hotel
Designed/Completed: 1996/1999
Location: Tiny-Lujiazui Area, East of Huangpu River, Shanghai
Total Floor Area: 110,000m²
Total Height/Floor: 40m/ 11 Floors
Type of Structure: Framed & Steel Structure

常州中华恐龙馆

Chinese Dinosaur Museum, Changzhou

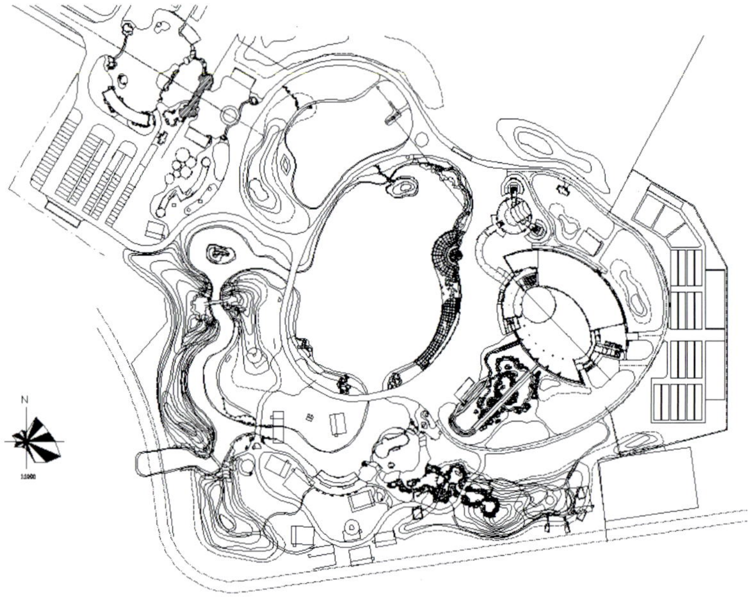

常州中华恐龙馆位于常州市开发区中华恐龙园，东邻沪宁高速公路，是一座结合高科技娱乐的恐龙主题自然博物馆，同时以动感的仿生造型组合博览、影视、游乐的建筑内容。建筑从侏罗纪化石鹦鹉螺的形态发展出一个以曲线的扇形组合的平面布局，空间造型上以恐龙动态和侏罗纪生物的皮骨羽翼为元素进行分解重构，塑造了一个充满张力和视觉冲击力的建筑造型，同时以龙形塔体传达了中国传统龙珠之戏的象征意念。

Chinese Dinosaur Museum Changzhou is located off Shanghai-Ningbo Highway and designed as a theme museum with hi-tech amusement facilities and motional bionic models incorporating the functions like museum, cinema and entertainment. The architectural appearance is inspired from the shape of the fossil nautilus of Jurassic Period and developed into a fan-shaped curving plan, yet the expressive form is created by disintegration and reconstruction of the dynamic states of dinosaur and skins, bones, feathers and wings of living things during Jurassic Period, reinforcing the basic themes of the museum. The expressiveness, which is reflected in the dragon-shaped tower, refers to the traditional significance of Two Dragons Playing with a Pearl.

建 设 单 位：常州新区管委会
功 能 用 途：科普、游乐
设计 / 竣工年份：1998 年 /2000 年
建 设 地 点：常州市开发区
总 建 筑 面 积 ：15000m^2
总建筑高度 / 层数：70m/3 层
结 构 形 式：框架结构、钢结构

Client: Changzhou New District Administrative Committee
Function: Popular Science, Amusement Facilities
Designed/Completed: 1998/2000
Location: Changzhou Development Zone, Jiangsu
Total Floor Area: 15,000m^2
Total Height/Floor: 70m/ 3 Floors
Type of Structure: Framed structure & Steel Structure

山东大厦

Shandong Building, Shandong

山东大厦是集宾馆、地市会议厅、大会堂为一体的高级民用建筑，面积为12.5万m^2。宾馆25层，690个客房，有普通客房、公务客房、豪华客房及总统客房。设有地市厅17个，以及800座会议厅和200座的国际会议厅各一个。2500座的省级大会堂有三层观众厅，除满足会议要求外，可满足交响乐、大型歌舞等演出要求。

The Shandong Building consists of a 5-star atrium hotel, prefecture/city level halls, enterprise level halls and a People's Hall. The 690-room, 25-storey hotel includes the junior suite, executive suite, deluxe suite and presidential suite. The 17 prefecture/city level halls host an 800-seat large convention hall, a 200-seat international convention center and other meeting venues of various size. The 2500-seat convention hall serves not only government functions but also business conventions and cultural performances.

建 设 单 位：山东大厦筹建处
功 能 用 途：宾馆、会议
设计 / 竣工年份：1993 年 /2002 年
建 设 地 点：山东济南南郊马鞍山路 2 号
总 建 筑 面 积 ：125000m^2
总建筑高度 / 层数：98m/25 层
结 构 形 式：框剪结构
合作设计单位：美国波特曼设计事务所

Client: Task Force of Shandong Building
Function: Convention Hall, Hotel
Designed/Completed: 1993/2002
Location: No. 2 Ma an Shan Lu, South Suburb, Jinan
Total Floor Area: 125,000m^2
Total Height/Floor: 98m/ 25 Floors
Type of Structure: Framed Shear Wall Structure
In association with John Portman & Associates. Inc.

杭州萧山国际机场

Hangzhou Xiaoshan International Airport, Hangzhou

杭州萧山国际机场坐落在杭州市萧山区东侧，是国内干线机场和重要的旅游机场，建筑面积 10 万 m^2，机场总体布局采用进港大道到达航站楼端头式布置，有 12 个停机位，配置 11 条全透明的固定登机桥，年旅客吞吐量至 2005 年达到 800 万人次。

航站桥高架上面连续漂动的雨篷和波浪起伏的屋顶象征着西湖水波纹和连绵起伏的西山群影。

Responding to Hangzhou's picturesque setting with its nearby lake and mountains, the design creates the impression of waves with the metal-clad roofs of the main hall, departure lounge and octagonal hub containing passenger lounges and VIP facilities. Layout of interior spaces is guided by a need for openness and clarity, achieved by means of large expanses of windows and clerestories, and a layout which maximizes operational efficiency and orientation. Landscaping is also a major design element: the design features a large pond, numerous trees and a water feature spanned by walkways between the terminal and parking lot. The project includes 12 parking aprons and 11 transparent passenger bridges. The airport handles eight million passengers in 2005.

建 设 单 位：杭州萧山国际机场工程指挥部
功 能 用 途：航空交通
设计 / 竣工年份：1998 年 /2000 年
建 设 地 点：杭州萧山区坎山镇
总 建 筑 面 积 ：100000m^2
总建筑高度 / 层数：23m/2～3 层
结 构 形 式：框架、钢结构
合作设计单位：加拿大 B+H 建筑设计事务所

Client: Hangzhou Xiaoshan International Airport Construction Headquarter
Function: Airport
Designed/Completed: 1998/2000
Location: Kanshan Town, Xiaoshan, Hangzhou
Total Floor Area: 100,000m^2
Total Height/Floor: 23m/ 2～3 Floors
Type of Structure: Framed & Steel Structure
In association with Bregman + Hamann Architects

浙江世界贸易中心

Zhejiang World Trade Center, Hangzhou

本工程位于宝石山下，黄龙洞旁，系西湖风景区的外围保护地带。主体建筑以“S”型曲线平面布局，采取高低错落台阶式的造型，由于整个建筑群体在平面、竖向三度空间均为曲面，使建筑物的立面轮廓俊秀、挺拔、错落有致，远处望去，颇有“黄龙腾飞”之感。建筑主体由宾馆、展览厅、写字楼三部分组成，在广场下面设有近9000m^{2}的地下停车场。该项目已成为我省及杭州市的国际贸易、国际展览和对外综合服务中心。

Located at the foot of Jade Mountain and adjacent to the Dragon Cave, the S-shaped main building comprises the hotel, exhibition hall and office building and is designed as a “terraced” structure . Under the concourse is occupied by a 9,000 square meter basement car parking area. With a curved free form the expressiveness, which is reflected in the prominent silhouettes is graceful and straightforward. From far away it conveys a sense of “dragon soaring into the sky”. Upon completion, it will host the international trade, international exhibition and comprehensive services to the public.

建 设 单 位：浙江世界贸易中心有限公司
功 能 用 途：宾馆、展览、写字楼
设计 / 竣工年份：1989 年 /1997 年
建 设 地 点：杭州市曙光路杭大路交叉口
总 建 筑 面 积 ：106830m^2
总建筑高度 / 层数：72m/18 层
结 构 形 式：框剪结构
合作设计单位：杭州市城建设计院

Client: Zhejiang World Trade Center Co., Ltd.
Function: Hotel, Exhibition, Office
Designed/Completed: 1989/1997
Location: Intersection of Shuguang Lu & Hangda Lu, Hangzhou
Total Floor Area: 106,830m^2
Total Height/Floor: 72m/ 18 Floors
Type of Structure: Framed Shear Wall Structure
In association with Hangzhou Urban Building Design Institute.

义乌市旧城改造暨市民广场建设规划

Yiwu Historical City Revitalization & Citizen's Plaza Master Plan, Zhejiang

该项目规划区域位于义乌市老城区，是义乌市的政治、文化、商业中心。规划结构为“一个中心，两条轴线”。用地内的道路网骨架为“Y”型，分为12个分区，功能为广场、公园、商贸、办公、商住等。在市民广场周围布置了市政府、大剧院、百货大楼、世纪商城等重要公建。规划设计顺承了当地人文及历史文脉，并注重古建筑与历史老街的保护。

该规划区内的大部分重要单体项目的设计均由我院中标实施并已完工。该规划获得方案国际招标一等奖，浙江省计划委员会优秀项目一等奖、省科技委员会优秀项目二等奖。

Respectful of the indigenous architecture and urban heritage, the conceptual design for the project is also in keeping with the principles of preservation of historical buildings and neighborhood. Predominantly located in the historical neighborhood as the political, cultural and economical hub of Yiwu, the project includes two main components: a center (Citizen's Plaza) and two axes. Roads within the site are meticulously outlined as Y-shaped layout. The site is broken down into 12 major areas such as the square, park, business, office and commercial residence, etc. Around the Citizen's plaza are taken up by municipal government, theatre, department store and century mall, etc. ZIAD has won the first prize in an international competition for the preliminary concept design and has been trusted with the designs for many major components that have been constructed and completed.

The master planning design for the project has won the First Prize of Excellent Design issued by Zhejiang Economic Planning Committee and the Second Prize of Excellent Design issued by Zhejiang Scientific and Technological Committee.

建 设 单 位：义乌市旧城改造建设指挥部
功 能 用 途：综合
设计 / 竣工年份：1999 年 /2003 年
建 设 地 点：浙江省义乌市老城区
总 建 筑 面 积 ：830000m²
总建筑高度 / 层数：99.9m/26 层
结 构 形 式：框架结构

Client: Yiwu Historical City Revitalization Headquarter
Function: Multi-use
Designed/Completed: 1999/2003
Location: Historical City Yiwu, Zhejiang
Total Floor Area: 830,000m²
Total Height/Floor: 99.90m/ 26 Floors
Type of Structure: Framed Structure

上海金钟广场

Golden Bell Plaza, Shanghai

金钟广场位于上海市淮海中路东段，龙门路口。广场体现了一种新形态、高品位、多凝聚的设计。它的特点之一是采用多体块组合的手法，在体块上运用八边形、半圆形、长方形等多体块重组，使之达到有机的整体；特点之二是丰富的天际轮廓线，建筑吸取上海传统建筑的丰富形式语汇，在加入城市天际线时，丰富和完善城市风貌；特点之三是在顶部和墙面细部处理上运用上海建筑的传统符号内涵。特别是在大厦顶部设计了一座“金钟塔楼”以体现“金钟广场”的象征意义。另外，在建筑外部采用多层次的泛光照明，使大厦在夜间也闪耀着光芒。

The architecture and planning of Shanghai Golden Bell Plaza is unique in its form, skillful in their execution and high standard of finishes. One of its main features in the design is well- defined by the combination of geometric elements on the facades like octagon, semicircle and rectangle. The second is reflected on its varied silhouettes that respect the symbolic traditional architecture in Shanghai and complement the city skyline. The third one is the application of the traditional Shanghai architectural symbol on the top and wall details, especially provided with a Golden Bell Tower on the top of the building in meaning of bringing good luck and fortune. In addition, flash lighting of various kinds is designed to illuminate the facades of the building at night.

建 设 单 位：上海金钟商业发展有限公司
功 能 用 途：商场、办公、娱乐、文娱
设计 / 竣工年份：1994 年 /1999 年
建 设 地 点：上海淮海中路东段龙门路口
总 建 筑 面 积 ：92000m²
总建筑高度 / 层数：208m/46 层
结 构 形 式：框剪结构

Client: Shanghai Jinzhong Commercial Development Co., Ltd.
Function: Retail, Office, Entertainment, Culture
Designed/Completed: 1994/1999
Location: Huai Hai Zhong Lu, Shanghai
Total Floor Area: 92,000m²
Total Height/Floor: 208m/ 46 Floors
Type of Structure: Framed Shear Wall Structure

杭州望湖时代广场—东方金座

East King • Lakeview Times Square, Hangzhou

建 设 单 位：杭州金瑞房产公司
功 能 用 途：办公、会议、娱乐、餐饮
设计 / 竣工年份：2002 年 /2006 年
建 设 地 点：杭州市岳王路
总 建 筑 面 积 ：53000m²
总建筑高度 / 层数：80m/23 层
结 构 形 式：框剪结构

Client: Hangzhou Jinrui Real Estate Company
Function: Office, Convention, Entertainment, Restaurant
Designed/Completed: 2002/2006
Location: Yuewang Lu, Hangzhou
Total Floor Area: 53,000m²
Total Height/Floor: 80m/ 23 Floors
Type of Structure: Framed Shear Wall Structure

东方金座位于杭州西湖东侧，由二幢23层的高层和4层裙房组成，并设有二层地下汽车库。高层北楼为大开间写字楼，南楼为SOHO小型办公用房，裙房设有会议、娱乐、餐饮、健身等多种配套服务功能。

为确保高层写字楼拥有最佳的西湖风光视觉景观，主楼在平面关系上呈现面向西侧的环抱状，裙房部分也在西侧形成一个较大规模的半圆形入口广场。南北高层之间留有较大的空间，避免了在西湖的沿湖景观上造成封闭感。建筑的外观造型设计采用了典型的现代建筑风格，设计中进行了多处圆弧形45°转角的细节化处理，增强了建筑的现代化和科技化气息。

Located at the east side of scenic Westlake, the complex is comprised of two 23-floor high-rise buildings, a 4-floor podium and 2-floor basement car parking. The north wing of the skyscraper is dominated by the open offices and the south wing accommodates the SOHO offices. Podium structure is designated by supporting functions as convention, entertainment, restaurant and health club facilities, etc. Its crescent-shaped plan of the main building embraces the lake with open arms and maximizes the magnificent views towards the Westlake. A grand semi-circular entrance plaza dominates the west side of the podium base. Between two high-rise slices lies an open space to minimize the "wall effect" to the lake side. The appearance of the complex is designed in a typical modern style and composed of circular and angular geometrical elements in details.

杭州西湖国贸中心

Westlake National Trade Center, Hangzhou

建 筑 单 位：杭州鸿瑛房地产开发公司
功 能 用 途：办公、商业、餐饮、娱乐
设计 / 竣工年份：2001 年 /2004 年
建 设 地 点：杭州市上城区湖滨路庆春路交叉口
总 建 筑 面 积：53088m²
总建筑高度 / 层数：23.9m/7 层
结 构 形 式：框架结构

Client: Hangzhou Hongying Real Estate Corp.
Function: Office, Retail, Restaurant, Entertainment
Designed/Completed: 2001/2004
Location: Intersection of Hubin Lu & Qingchun Lu,Shangcheng District, Hangzhou
Total Floor Area: 53,088m²
Total Height/Floor: 23.90m/ 7 Floors
Type of Structure: Framed Structure

西湖国贸中心位于杭州市湖滨1号地块，庆春路与湖滨路的交叉口，直接面向西湖。建筑的主要功能为高档综合写字楼、商场、娱乐、餐饮等。建筑为地上7层，地下2层。在总体布置上，写字楼主体为"L"形的布局及面向西湖的弧形退台布置，保证了大部分写字间都能面向西侧的西湖和保俶塔。建筑物距道路交叉口退出较大的空间，形成一个以古钱塘门遗址为中心的下沉式广场，使地下层的娱乐、餐饮场所能够自然采光，也有了与室外相通的出入口。

Hangzhou Westlake National Trade Center is a mixed-use development composed of commercial, office, restaurant and entertainment functions with 7 floors aboveground and 2 basement levels. The L-shaped office building are set back in circular form, which maximizes the numbers of the office facing the West Lake at the west and enjoys a fantastic view over Baoshu Pagoda. Extensive space to be reserved away from the intersection creates a sunken plaza occupied by an ancient Qian-tang Gate relic, in which the underground entertainment functions and restaurants are naturally lit and accessible to the outside.

杭州雷峰塔景区

Leifeng Pagoda Resort, Hangzhou

该景区位于杭州西湖南端夕照山，北面毗邻西湖，南接南屏山，周边景点丰富，自然环境可谓得天独厚。景区建筑突出雷峰塔为景区主题，主要由塔区、塔院、山门与内院、购物街、如意苑、翠芳园、碑亭、望湖楼等组成，整个景区建筑群，依山傍势，紧密结合，位置恰当，互为依托，高低错落，疏密有度，道路分区明确。建筑造型结合了江南园林建筑和江南民居建筑的特点，丰富了整个景区的景观效果，使该景区成为西子湖畔又一旅游胜地。

Situated on the Sunset Glow Hill near Westlake and bounded Nanping Hill at the south, the project is richly endowed by nature surrounded by a rich cultural and environmental heritage of the City of Hangzhou. Leifeng Pagoda is designed as the theme of the resort that consists of the Pagoda, Pagoda Court, Gate, Inner Court, Pedestrian Street, Good Luck Garden, Green Fragrant Garden, Stone Tablet Pavilion and Lake Viewing Pavilion. Due to its splendid location of the site with animated hills at its back and valleys pond between, the various components are located properly to blend into the terrain of the site and complemented each other by their terraced structures and well-designed density. Roads within the site are meticulously designed and well defined. Incorporating the spirit of the architecture of garden and the indigenous residence in South of China, the form of the project enriches the landscape of the site that creates another scenic spot to the city of Hangzhou.

建 设 单 位：杭州雷峰塔旅游文化发展有限公司
功 能 用 途：餐饮、旅游、管理、陈列
设计 / 竣工年份：2000 年 /2003 年
建 设 地 点：杭州西湖南端夕照山上
总 建 筑 面 积 ：7000m²
总建筑高度 / 层数：2 层
结 构 形 式：框架结构

Client: Hangzhou Leifeng Pagoda Tour Cultural Development Co., Ltd.
Function: Restaurant, Tour, Administration, Display
Designed/Completed: 2000/2003
Location: South of Westlake, Hangzhou
Total Floor Area: 7,000m²
Total Height/Floor: 2 Floors
Type of Structure: Framed Structure

杭州嘉德广场

Jiade Plaza, Hangzhou

建 设 单 位：浙江嘉德房地产发展有限公司
功 能 用 途：办公、商业
设计 / 竣工年份：2001 年 /2003 年
建 设 地 点：杭州市庆春路交叉口
总 建 筑 面 积 ：60460m²
总建筑高度 / 层数：100m/32 层
结 构 形 式：框剪结构

Client: Zhejiang Jiade Real Estate Development Co. Ltd.
Function: Office Building, Commercial Facilities
Designed/Completed: 2001/2003
Location: Qingchun Lu, Hangzhou
Total Floor Area: 60,460m²
Total Height/Floor: 100m/ 32 Floors
Type of Structure: Framed Shear Wall Structure

嘉德广场位于杭州市庆春路与中河路交叉口，东邻工商银行大楼及中河绿化带、庆余亭，北侧为农业银行大楼。它由 27 层主楼和 5 层裙房、3 层地下室组成，基地面积 8260m²。是一座集现代商场、甲级写字楼为一体的大型综合性建筑，建筑体形为长方形与椭圆形有机穿插，融为一体，赋予建筑双广场与垂直三重景观，并与中河绿化带交相辉映。

建筑外部轮廓纯净，内部空间丰富，观光电梯、踏步、采光楼梯及阳光充沛的屋顶花园都体现出现代建筑通透、开敞、多元的内涵。

Jiade Plaza is situated at the intersection of Qingchun Lu and Zhonghe Lu, bounded by Hangzhou Industrial & Commercial Bank, the city green belt and Qingyu Pavilion to the east. The Agricultural Bank is to the north. The project occupies an overall site area of 8,260 sq.m and consists of a 27-floor main building, a 5-floor podium structure and 3 levels underground. The 100-meter-high complex contains state-of-the-art retail and office space of Grade A. The profiles of the project are composed of rectangular and oval-shaped volume in an organic displacement that offers the building a duplex plaza and triplex vertical landscaping, especially complements to the immediate context.

The design suggests an austere, rectangular building closed to the outside and intended to achieve spatial diversity and formal richness only in the interior. The openness, transparency and diversity of the modern architecture are endowed to the project by the configuration of prospective elevator, spacious steps, skylighted staircase and sunshined roof garden.

嘉德廣場

淮安淮海第一城

No.1 City of Huai Hai Huai An, Jiangsu

“淮海第一城”位于淮安市中心核心商圈清河区内，占地近300亩，北区为高级住宅区，南区为集中组团式商业广场及住宅区。其中住宅从18层至24层不等、高低错落，曲折变化，形成灵动的城市天际轮廓线。商业及20层酒店、20层写字楼、公寓办公区布置在淮海路边，裙房则布置最利销售并符合当地商业习惯的小开间精品店，层数以一、二层为主。双塔楼的布置力求作为淮海路上的重叠高音符号，为“淮海第一城”树立鲜明形象，并有效提升了城市品位。在地块中心，有大型广场、下沉庭院，还有流畅曲线形成的退台空间，这样的布局带动了人气，提高了商业旺铺的价值，同时也巧妙地利用了地下空间。

Located in the downtown of Huai′an, the development occupies a site area of 20 hectares including a high-grade residential area at the north and a clustered commercial plaza and residences at the south. The residential buildings, from 18 to 24 floors, are staggered in height providing a distinctive and dynamic skyline symbol. Along the boulevard of Huaihai Road are dominated by the commercial plaza, 20-floor hotel, 20-floor office building and SOHO office building. Podium structure is taken up by one or two-storey small open boutiques. The twin tower of the main building structure forms an ensemble in a modern international architectural style ,futuristic in shape.The central area of the site, provided with grand concourse, sunken courtyard and curve-shaped setback space, attracts life and response to the activities of the project, while maximizes the use of the underground level.

建 设 单 位：淮安华水曙光置业有限公司
功 能 用 途：住宅、办公、商业
设计 / 竣工年份：2006 年 / 在建
建 设 地 点：淮安市清河区
总 建 筑 面 积 ：400000m^2
总建筑高度 / 层数：95m/22 层
结 构 形 式：框架结构

Client: Huai' an Holley-Shuguang Real Estate Co., Ltd.
Function: Residence, Office, and Commercial
Designed/Completed: 2006/under construction
Location: Qinghe District, Huai' an, Jiangsu
Total Floor Area: 400,000m^2
Total Height/Floor: 95m/22 Floors
Type of Structure: Framed Structure

长沙中天广场

Zhongtian Plaza, Changsha

长沙中天广场位于长沙市最繁华的市中心地带，五一西路4号、6号地块，总用地面积15125m²。设计将建筑整合为前后两组。前区公建由两幢塔式建筑组成，写字楼、酒店式公寓各占东西一角，两组建筑内侧为弧形，通过上部连接，沿街立面呈门字形。东侧写字楼边上设突破天际线的通高玻璃塔形成标志性，而门字形建筑与周边其他百米左右的塔式高层又形成视觉上的对比，突出了此地块的个性表现。公建下面部分为4层，屋面上设有绿化，可作为商业建筑的休闲辅助设施，同时又可作为后区高层住宅的花园，前后两组建筑呈围合形，相互间关系紧凑又保持其相对独立性，建筑群体具有独特的个性。

Situated in the bustling center of downtown, the development occupies a site area of 15,125 square meters. It includes three main components: an office building, a service apartment and residential buildings. The office building and the service apartment separately dominates the east and west corner in front of the residential complex. With their curving inner sides and connected on the upper level, such composition appears like an "Entry Gate" on the street level . A full height glazed tower is attached to the office building and pierces into the skyline as a symbol. The "Gate" building visually competes with the other towers not far away that accentuates the personality of the site. On the roof of the 4-storey podium base is provided with a landscaping garden offering a magnificent view to the residents. The architectural composition makes the components closely connected but also independent.

建设单位：湖南天翔房地产有限公司
功能用途：办公、酒店式公寓、住宅
设计/竣工年份：2004年/在建
建设地点：湖南长沙五一西路
总建筑面积：170000m²
总建筑高度/层数：100m/33层
结构形式：框架核心筒结构

Client: Hunan Tianxiang Real Estate Development Co., Ltd.
Function: Office, Service Apartment, Residence
Designed/Completed: 2004/under construction
Location: Wuyi Lu（West),Changsha, Hunan
Total Floor Area: 170,000m²
Total Height/Floor: 100m/33 Floors
Type of Structure: RC Framed Core-wall Structure

杭州福雷德广场

FRID Plaza, Hangzhou

建 设 单 位：澳大利亚 FRID 投资发展有限公司
功 能 用 途：综合
设计 / 竣工年份：2003 年 / 在建
建 设 地 点：杭州下沙
总 建 筑 面 积 ：190000m²
总建筑高度 / 层数：100m/26 层
结 构 形 式：框架结构
合作设计单位：澳大利亚 HC 建筑设计公司

Client: FRID (Hangzhou) Investment and Development PTY LTD.
Function: Multi-used Complex
Designed/Completed: 2003/under construction
Location: Xiasha Economic & Technologic Development Zone, Hangzhou
Total Floor Area: 190,000m²
Total Height/Floor: 100m/ 26 Floors
Type of Structure: RC Framed Structure
In association with HC Consultant Pty. Ltd.

杭州福雷德广场由澳大利亚FRID投资发展有限公司投资开发，福雷德(杭州)置业发展有限公司承建。项目位于杭州下沙经济技术开发区和高教园区的中心地段，总用地55200m²，由会展中心、购物中心、配套公寓三大功能区块组成。项目一期包括4幢高层公寓，约600多套住房，底层设置沿街商铺，中央景观绿地近6500m²，小区内实现人车分流。二期由高达100m的酒店式公寓、大型会展中心和将近4.5万m²的大型商业中心SHOPPING MALL组成，是一个集生活、居住、休闲、娱乐和商业为一体的社区。简洁的造型，线条明快、色彩柔和的外立面，共同营造出广场大气、优雅的气度。

The project is situated in the center of Hangzhou Xiasha Economic & Technologic Development Zone and High Education Park. It occupies an overall site area of 55,200 square meters including the convention and exhibition center, shopping mall and condominium building with 53,000-square-meter basement level. Phase One contains 4 high-rise condominiums of more than 600 units, street-front shops on the ground floor and the central landscaping area of 6,500 square meters around. Phase Two consists of a 100-meter-high hotel apartment, convention and exhibition center and a shopping mall of approximate 45,000 square meters. The simple form of the building and the formal texture of the facades tinted with soft colors create a grandiose and graceful image to the project.

杭州运河广场

Grand Canal Plaza, Hangzhou

建 设 单 位：杭州市拱宸桥旧城改造指挥部
功 能 用 途：博物馆、广场
设计/竣工年份：2001 年 /2005 年
建 设 地 点：杭州市拱宸桥西
总 建 筑 面 积：45000m²
总建筑高度/层数：15m/2 层
结 构 形 式：框架结构

Client: Hangzhou Gongcheng Bridge Old City Revitalization Headquarter
Function: Museum, Plaza
Designed/Completed: 2001/2005
Location: West of Gongchen Bridge, Hangzhou
Total Floor Area:45,000m²
Total Height/Floor: 15m/ 2 Floors
Type of Structure: Framed Structure

杭州运河广场上的主要建筑是杭州运河博物馆。杭州运河博物馆位于京杭大运河古桥－拱宸桥的西面，由于"拱宸桥"自身是运河博物馆的最重要的室外现存文物，所以运河广场的东西轴线是以烘托突出拱宸桥为目的，从东到西布置了一系列的水景，使人们联想到运河。水景、牌坊、桥头、桥、运河构成了运河广场的主景。南北轴线把区政府、中心广场和博物馆三者有机组合在一起。运河广场布置有集会广场、表演台、休闲林、座椅、喷泉、雕塑等。博物馆环抱广场呈扇形布置，造型上采取平、坡结合，主入口及两边的展厅主要空间采取两坡顶，半悬厅上采用圆形坡屋顶，其余部位以平屋顶为主，烘托着具有东方传统韵味的博物馆。

Since Gongchen Bridge itself is renowned as the most important outdoor existing relic of Grand Canal Museum, the conceptual design for the Grand Canal Plaza aims to reflect the significance of Gongchen Bridge at its west-east axis. A series of water elements are thus introduced throughout the east and the west reminiscent of the Grand Canal. Inclusion of water features, gateway, bridge tower, bridge and canal acknowledges the main landscaping features within the site. The link between the south and north axis is the organic placement of Prefecture Government Building, Central Plaza and Grand Canal Museum. The development contains the concourse, performance stage, woods for relaxation, benches, fountain and sculptures, etc. The fan-shaped museum encircles the plaza. Double-pitched roofs are provided to the main entry and exhibition halls located at the two ends, yet above the semi-suspended lobby uses the round-pitched roof and the others are covered by the flat roofs. This is to recognize the spirit and significance of the traditional vernacular architecture in the design of the museum.

杭州广复大厦

Guangfu Mansion, Hangzhou

这是一幢特色鲜明的大楼。主楼由26层和19层两幢弧形写字楼组成，坐落在5层圆弧形商业裙房之上。建筑立面处理简洁大方，采用石材和有韵律感的点窗相结合，取得了较好的效果。沿着周边道路上看这幢弧形大楼的外形，随着位置的移动具有不断变化的轮廓，给人印象深刻。主楼顶部采用简洁的几何形体处理手法设计了弧形玻璃拱顶，不仅造型美观，也有利于植物生长。主楼顶部屋顶花园与裙房逐层退台的绿化平台相呼应，形成这幢建筑物丰富多彩的第五立面。

The main building includes two crescent-shaped office buildings of 26 stories and 19 stories located above a 5-storey fan-shaped podium structure. The simple form of the building is decorated with stone materials and transparent spots. The skeletal form of the tower is various when it is visible from the immediate that is impressive to the visitors. A glazed dome of simple form is designed to the top of the main building. It is expressive in the form but also benefit to the growth of the plants. Roof garden on the top level of the main building is provided corresponding to the stepped down landscaped terrace of the podium base.

建 设 单 位：杭州广宇集团股份有限公司
功 能 用 途：商业、办公
设计 / 竣工年份：2001 年 /2003 年
建 设 地 点：杭州城站广场
总 建 筑 面 积 ：70778m²
总建筑高度 / 层数：99.3m/26 层
结 构 形 式：框肢剪力墙结构

Client: Hangzhou Cosmos Group Co., Ltd.
Function: Commercial, Office
Designed/Completed: 2001/2003
Location: Railway Station Plaza, Hangzhou
Total Floor Area: 70,778m²
Total Height/Floor: 99.30m/ 26 Floors
Type of Structure: RC Shear Wall Support on Frame Structure

上海宝安大厦

Bao' an Tower, Shanghai

建 设 单 位：上海宝安企业有限公司
功 能 用 途：写字楼、商务、酒店
设计/竣工年份：1996年/2003年
建 设 地 点：上海市浦东区
总 建 筑 面 积：89260m²
总建筑高度/层数：138m/38层
结 构 形 式：框筒结构
合作设计单位：浙江大学建筑设计研究院

Client: Shanghai Bao' an Industrial Co., Ltd.
Function: Office, Business, Hotel
Designed/Completed: 1996/2003
Location: Pudong New Area, Shanghai
Total Floor Area: 89,260m²
Total Height/Floor: 138m/ 38 Floors
Type of Structure: Framed Core-wall Structure
In association with Zhejiang University Building Design & Research Institute

宝安大厦位于浦东新区中心崂山东路，是浦东最早的高层建筑之一。该大厦充分利用三角形地块，以切角三角形构成呼应地形的建筑体块，周边布置环形通道和绿化带，交通流畅便捷，工作分区合理。大厦是一座集写字楼、酒店、餐饮、娱乐、购物等为一体的多功能综合建筑，也是一座个性突出的城市综合体建筑，其丰富的形态和空间变化，为浦东的城市空间和城市轮廓线作出了积极的贡献。本项目2005年荣获浙江省优秀设计一等奖。

Located in a triangular-shaped site, the development fully utilizes the assets. A canted triangular form of structure is introduced to meld into the site. Around the tower is distributed by the loop road and the plants. The overall plan of the complex excels with an efficient transport link for users and with the high flexibility of all facilities. Incorporating the comprehensive program of office, hotel, restaurant and retail, the mix-used building has a distinctive presence. The design intends to achieve spatial diversity and formal richness that will prove to make contribution to the city' s context and the skyline of Pudong Shanghai. The project has awarded the honor of First Prize of Excellent Design Zhejiang 2005.

柳泉正承商务广场

Liuquan · Landsun Plaza, Shandong

柳泉正承广场位于淄博市淄川区，占地面积6hm²。项目包含休闲广场、购物中心、酒店式公寓以及办公等多种功能。广场设计根据地形地貌，结合河道的天然景观资源，创造出富有立体感的亲水、亲自然的现代园林化市民休闲场所。广场北端建有3幢高层酒店式公寓，共160套，1幢需进行立面改造的保留高层办公楼。广场地下一层建造3万m^2的大型超市，地下二层建造15000m^2，近300个泊车位的大型停车场。建成后的柳泉正承商务广场将成为淄川的标志性城市空间与建筑，成为淄川最具现代化的休闲娱乐购物中心。

Incorporating the comprehensive program of entertainment, retail, restaurant, service apartment and office functions, the development occupies a site area of 6 hectares. Melding into the terrain of the site and integrating into the natural landscape of the river nearby, the design for the plaza creates a modern gardening architecture that is in close proximity to the water and the nature. Three high-rise service apartments of 160 units are located at the north end of the plaza. An old office tower is to be renovated to its facades. A grand supermarket of 30,000 square meters is provided on the first basement level. On the second basement floor is occupied by a car parking of 15,000 sq.m for 300 cars. The Plaza would be a signature building of the city and the most state-of-art facility to the local citizen upon completion.

建设单位：山东省南郊集团
功能用途：综合
设计/竣工年份：2004年/在建
建设地点：山东省淄博市淄川区
总建筑面积：100000m²
总建筑高度/层数：72m/24层
结构形式：框架结构

Client: Shandong South Suburb Group
Function: Multi-used
Designed/Completed: 2004/under construction
Location: Zichuan District, Zibo, Shandong
Total Floor Area: 100,000m²
Total Height/Floor: 72m/ 24 Floors
Type of Structure: RC Framed Structure

杭州标力广场

Triplenic Plaza, Hangzhou

建 设 单 位：标力（杭州）地产有限公司
功 能 用 途：办公、商场、金融
设计 / 竣工年份：1996 年 /2000 年
建 设 地 点：杭州延安北路
总 建 筑 面 积 ：100500m²
总建筑高度 / 层数：99.60m/27 层
结 构 形 式：框筒结构

Client: Triplenic (Hangzhou) Real Estate Co., Ltd.
Function: Office, Retail
Designed/Completed: 1996/2000
Location: Yan' an Bei Lu, Hangzhou
Total Floor Area: 100,500m²
Total Height/Floor: 99.60m/ 27 Floors
Type of Structure: Framed CoreWall Structure

本工程是杭州市中心区的高密度、高容积率建筑之一，建筑密度50.6%,建筑容积率6.62。建筑由8层的商业大厦，15层的金融中心和27层的办公楼组合而成高层建筑群，建筑群的每个功能个体均有独立的出入口和各自的垂直交通中心。建筑群的立面统一设计，外墙均为干挂石材，建筑形体敦厚简洁又颇具现代感。

Located in the downtown of the city, the development includes an 8-storey commercial building, a 15-storey financial center and a 27-storey office tower. Each component is provided with dedicated access and separated vertical circulation. The functions have similar facades of stone materials to exhibit their modern and sophisticated simplicity.

杭州武林门旅游客运中心

Tourism Center Wulinmen, Hangzhou

建 设 单 位：杭州坤和旅游客运中心有限公司
功 能 用 途：旅游码头、商场、写字楼
设计 / 竣工年份：2003 年 / 在建
建 设 地 点：杭州武林门
总 建 筑 面 积 ：120000m^2
总建筑高度 / 层数：115m/31 层 + 地下 3 层
结 构 形 式：框剪结构
合作设计单位：德国 GMP 国际建筑设计有限公司

Client: Hangzhou Canhigh Tourist Transfer Center
Function: Ferry Terminal, Retail, Office
Designed/Completed: 2003/under construction
Location: Wulinmen Square, Hangzhou
Total Floor Area: 120,000m^2
Total Height/Floor: 115m/ 31 Floors + 3 Floors Basement
Type of Structure: Framed Shear Wall Structure
In association with Architekten von Gerkan, Marg und Partner

本工程是集旅游码头、商场、写字楼等多功能为一体的综合性公共建筑。基地位于杭州市中心武林广场西北，著名的京杭大运河在其东北侧贴邻穿过，与杭州西湖文化广场隔河相望。31层高的主体建筑在基地东南端，高度约为115m，裙房为5～9层，与整个塔楼建筑有机的连为一整体，地下3层，总建筑面积约12万m^2。该建筑的造型具有强烈的雕塑风格，布局十分严谨，全部的柱网统一为一组模数，主体建筑锯齿形造型突出了它强有力的立体效果，整组建筑环绕而成的旅游码头前区广场与京杭大运河以及西湖文化广场建筑群的壮观景色遥相呼应，形成了杭州市区又一道亮丽的风景线。

The design refers to the historic meaning of Wulinmen Square as a connection between water and city. Additionally it develops the connection between the old southern town center, the new cultural center up north and the Jing-Hang Canal. The 5 to 9-storey podium structure is integrated into the 31-storey tower building. The design suggests an expressive sculptural form and proportional architectural composition. A unified module system is introduced to all column networks. Due to the comb-like and "terraced" structure of the complex the expressiveness is accentuated. The modern architecture corresponds to the magnificent and grandiose Grand Canal and the Westlake Cultural Center.

上虞市民广场

Civic Plaza, Shangyu

建 设 单 位：上虞市民中心建设指挥部
功 能 用 途：综合用房
设计 / 竣工年份：2003 年 / 预计 2008 年
建 设 地 点：上虞市城北新区
总 建 筑 面 积 ：238000m^2
总建筑高度 / 层数：78m/4～16 层
结 构 形 式：框架结构

Client: Shangyu Civic Plaza Construction Headquarters
Function: Multi-used
Designed/Completed: 2003/2008
Location: New District of North City, Shangyu
Total Floor Area: 238,000m^2
Total Height/Floor: 78m/ 4～16 Floors
Type of Structure: Framed Structure

上虞市市民中心位于上虞市城北曹娥江畔，是城北新区的标志性建筑群，是带动城市沿江向北发展的一个龙头。整个中心用地为38.66hm^2，由行政主楼（12层，10.5万m^2）、市民服务中心（5层，1.8万m^2），后勤服务中心（5层，1.6万m^2），档案馆（5层，1.3万m^2），图书馆（5层，1.2万m^2），文化艺术中心（5层，1.5万m^2），青少年宫（5层，1.5万m^2），会展中心（3层，1.5万m^2），地下停车场（3.2万m^2）等九个子项组成，工程分期实施，首期工程为青少年宫、图书馆文化艺术中心、档案馆、市民服务中心和后勤服务中心，整个工程预计于2008年全部完工。

Located near the bank of Cao' er River, the project is designed as an icon building in the New District of the city and intends to develop the city into the north along the river. It occupies a total site area of 38.66 hectares including nine components: 12-storey administrative main building of 105,000 sq.m., 5-storey civic service center of 18,000 sq.m., 5-storey back-of-house of 16,000 sq.m., 5-storey archives hall of 13,000 sq.m., 5-storey library of 12,000 sq.m., 5-storey culture and art center of 15,000 sq.m., 5-storey Youth's Palace of 15,000 sq.m. , 3-storey Convention and Exhibition Center of 15,000 sq.m. and underground car parking of 32,000 sq.m. The project will be divided into several phases and expected to be completed in 2008.

天辰国际广场

Tianchen International Plaza, Hangzhou

建 设 单 位：杭州泰和房地产开发有限公司
功 能 用 途：综合用房
设计/竣工年份：2004 年 /2006 年
建 设 地 点：杭州市萧山区
总 建 筑 面 积 ：$140000m^2$
总建筑高度/层数：109m/26 层
结 构 形 式：框架剪力墙结构

Client: Hangzhou Taihe Real Estate Development Co., Ltd.
Function: Multi-used
Designed/Completed: 2004/2006
Location: Xiaoshan, Hangzhou
Total Floor Area: $140,000m^2$
Total Height/Floor: 109m/ 26 Floors
Type of Structure: Framed Shear Wall Structure

该工程包括一幢办公楼、一幢酒店式办公楼、三幢住宅以及商场、娱乐、酒店等部分。南侧的三幢住宅处理成"L"形的形态，由此产生三个半围合的外部空间达到城市空间与内部空间内外交融的效果。南北两侧为外向型的广场，对应解决大型商场以及中小型店铺的出入口问题。中部为内向的庭院，小区住宅的步行主入口也是公寓区的形象入口。办公楼主楼则以点式的造型设于地块的最北端，前面设有绿化广场，成为其主入口。裙房的屋顶庭院形成一个主体的生态系统，成为人们休息、交流、观景的场所。在造型上，跌落式的屋顶花园富有节奏变化，避免了裙房与主楼脱节的单调组合方式，该建筑群成为沿市心北路的一道独特风景。

The project includes an office building, a service office, three residential buildings, retail, hotel and entertainment functions. At the south the architectural composition of three residential buildings is designed as a L-shaped configuration that provides three semi-enclosed courtyards melding the site and the architecture into a single, comprehensive expression. At the two ends of the north and south a plaza is located providing the entry to the retail store and small shops. A courtyard in the middle serves as the main entrance to the residences. The main office tower dominates the very northern end with a landscaped concourse before it as the main entrance. Designed as a primary ecological system, the roof garden on the top of the podium structure serves as a place for people to relax, communicate and enjoy the views. The stepped down rooftop garden integrates the podium with the main structure yet could have a distinctive presence along the main boulevard of the city.

上海浦东机场国航工作区

Shanghai Pudong International Airport • International Flight Operation Area,Shanghai

建设单位：中国国际航空公司上海基地
功能用途：办公、宾馆
设计/竣工年份：2004年/2007年
建设地点：上海浦东
总建筑面积：95000m²
总建筑高度/层数：26m/6层
结构形式：框架结构

Client: CA/CCA Shanghai Base
Function: Office, Hotel
Designed/Completed: 2004/2007
Location: Pudong, Shanghai
Total Floor Area: 95,000m²
Total Height/Floor: 26m/ 6 Floors
Type of Structure: Framed Structure

基地位于上海浦东新区机场大道南侧，总用地面积80452m²。规划总建筑面积95000m²，包括基地业务用房、机组过夜、旅客延误、单身宿舍、配餐、综合保障、库房、维修车间用房等内容。国航作业区的总平面采用了网络化的规划格局，简洁明了，具有较大的可调整性和灵活性。设计从建筑的性质、功能特征出发，以使用、管理便利为主，平面布置和室内空间力求规整、合理、现代化。建筑造型在强调简洁体块的穿插与体积感的同时，用金属、玻璃等现代材料细化立面风格，将绿地与水系引入基地，形成园区中心绿地，达到建筑和环境的和谐统一。

Occupying a total site area of 80,452 square meters, the development consists of offices, accommodation for the crew and the travelers, dormitory for the bachelor, pantry, multi security, storage and maintenance workshop, etc. The architectural composition shows a clear distinction between and within the network arrangement that provides great flexibility in the design. The conceptual design responds to the functional requirement of the development and recognizes the accessibility of the operation and the management. The appearance of the building is simple and clearly expressed through the articulation of massing and the use of materials such as the metal and glass. The inclusion of the nature-trees, water, and planting-permeates the complex and acknowledges the ecological design concept, providing a central landscaping area.

昆明文明路历史街区保护与改造规划

Master Plan for Wenming Lu Historical District Preservation & Rejuvenation, Kunming

文明路历史街区是近代昆明的政治和商业中心，人文历史遗址丰富，目前是全国现存规模最大的传统民居片区。规划充分考虑了未来多种经营的要求，以求达到社会效益与经济效益的双赢，并在整体上使传统风貌与现代城市相协调，使街区在完整保存传统历史风貌的基础上，成为具有浓郁地方特色，集历史文化展示、旅游观光、商贸、居住、休闲为一体的综合性街区。设计手法采用“五条线、四个板块”的分区保护、成片规划思路，确立文保和维修建筑的修缮方案，更新与重建部分传统民居，并在正义路沿线建造现代化商业设施，完善片区的城市功能。该项目已通过国务院紫线保护规划论证。

Wenming Lu Historical District is the political and business center of Kunming city in modern times, rich in historical cultural relics. It is the largest traditional vernacular housing district that exists in China at present. The design for the development anticipates expansion of business functions and attempts to win in the social and economical aspects, while incorporates the traditional vernacular into the modern city and creates a comprehensive development including historical cultural exhibition, housing, entertainment, trade commercial and tourism functions. The major conceptual components of the master plan include the principle of “five lines and four lots” that provides preservation to different districts but prepares overall master plan to the site, while sets up the concept for the preservation of cultural relics , renovation of the historical buildings and rejuvenation/reconstruction of some traditional vernacular houses. The state-of-the-art commercial facilities would be constructed along Zhengyi Lu to improve the urban functions in the district. The design for the development has been approved by the State Council under the Managing Urban Purple Line Review（Review to the Management of Historic Culture and Heritage).

建 设 单 位：昆明市之江置业有限公司
功 能 用 途：传统民居改造
设计 / 竣工年份：2003 年 / 在建
建 设 地 点：昆明市文明路
总 建 筑 面 积 ：205000m²
总建筑高度 / 层数：24m/5 层
结 构 形 式：框架结构

Client: Kunming Zhijiang Real Estate Co., Ltd.
Function: Traditional Housing Renovation
Designed/Completed: 2003/under construction
Location: Wenming Lu, Kunming
Total Floor Area: 205,000m²
Total Height/Floor: 24m/ 5 Floors
Type of Structure: RC Framed Structure

办公建筑
Office Building

浙江大学国家级科技园

National Scientific Park • Zhejiang University,Hangzhou

浙江大学科技园位于浙江大学玉泉校区西北侧，地块西北方向临西溪路，规划道路宽30m，地块内部起伏较大，场地呈狭长方形，南北窄、东西长，地块总用地面积33170m²。工程由6层主体条形建筑（包括研发、孵化、贸易、中介服务机构以及办公）联系其余各单体（包括沿街的休闲、会议、管理用房以及背山的三幢点式办公楼及公寓），并设有3个地下车库。地面以上总建筑面积49710m²，地面以下建筑面积 16280m²。

Located in the northwest of Zhejiang University Yuquan Campus, the project covers a total site area of 33,170 square meters with a total floor area of 49,710 square meters aboveground and 16,280 square meters below grade. It is composed of 6 rectangular-shaped main buildings including the functions of research and development, incubation of hi-tech enterprise, trade, agents and offices. The other individual buildings include the recreation, meeting, administrative office and three spot-typed office buildings and apartments. It has three basement levels for car parking and occupies 49,710 sq.m above ground and 16,280 sq.m below groude.

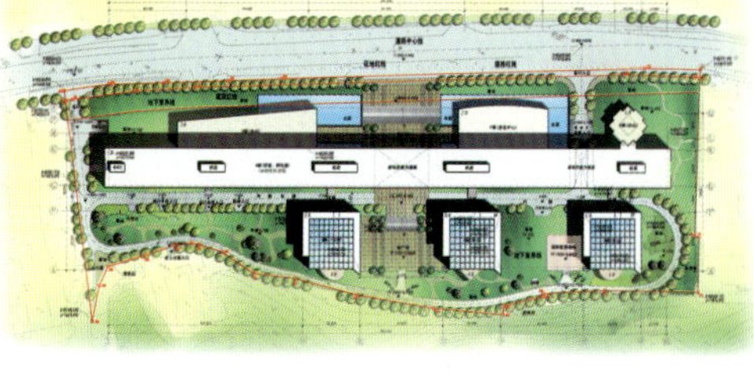

建 设 单 位：浙大科技园有限公司
功 能 用 途：科研、办公
设计 / 竣工年份：2002 年 /2003 年
建 设 地 点：杭州西溪路
总 建 筑 面 积 ：65990m²
总建筑高度 / 层数：35m/8 层
结 构 形 式：框架结构
合作设计单位：德国 HLD 设计事务所

Client: Zhejiang University Scientific Park Co., Ltd.
Function: Scientific Research,Office
Designed/Completed: 2002/2003
Location: Xixi Lu, Hangzhou
Total Floor Area: 65,990m²
Total Height/Floor: 35m/ 8 Floors
Type of Structure: Framed Structure
In association with HLD International Planungsgesellschaft mbH

CONFERENCE 会议中心

浙江省高级人民法院

Zhejiang Province High People's Court, Hangzhou

建 设 单 位：浙江省高级人民法院
功 能 用 途：办公、审判
设计 / 竣工年份：1998 年 /2000 年
建 设 地 点：杭州市马塍路 5 号
总 建 筑 面 积 ：23657m^2
总建筑高度 / 层数：70m/18 层
结 构 形 式：框剪结构

Client: Zhejiang Province High People's Court
Function: Office, Courtroom
Designed/Completed: 1998/2000
Location: No.5 Mateng Lu, Hangzhou
Total Floor Area: 23,657m^2
Total Height/Floor: 70m/ 18 Floors
Type of Structure: Framed Shear Wall Structure

浙江省高级人民法院位于杭州马塍路5号，建设项目包括审判办公大楼、法官培训中心及南配楼。主体建筑审判办公大楼以其纪念碑式的庄严形象和富有空间感和装饰性的细部表现了司法建筑的性格和法文化的思想内涵。建筑总占地 14518m^2。

The development is composed of three main functions: the courtroom and office building, training center for lawyers and the south-end back-of-the-house. It covers a total site area of 14,518 square meters. The courtroom and office building of monumental scale is designed as the main structure coupled with textured detail and formal richness to reveal the nature of the judicial architecture and the essence of law culture.

凤起·天翔时代大厦

Fengqi Tianxiang Times Tower, Hangzhou

建 设 单 位：浙江天翔房地产开发有限公司
功 能 用 途：商业、写字楼
设计 / 竣工年份：2002 年 /2005 年
建 设 地 点：杭州凤起路、新塘路交叉口
总 建 筑 面 积：40000m²
总建筑高度 / 层数：83.3m/23 层
结 构 形 式：框剪结构

Client: Zhejiang Tianxiang Real Estate Development Co., Ltd.
Function: Commercial Office Building
Designed/Completed: 2002/2005
Location: Southeast Corner of Fengqi Lu & Xintang Lu, Hangzhou
Total Floor Area: 40,000m²
Total Height/Floor: 83.30m/23 Floors
Type of Structure: Framed Shear Wall Structure

凤起·天翔时代大厦位于杭州凤起东路与新塘路交叉口西南角，由一幢23层高层写字楼与四层裙房组成，地下二层为停车，地下一层为商业用房。

该项目包括地下一层共有5层商业空间，设计采取了商业步行内街的处理手法，很好地解决了商业店面进深过大的不足，创造了丰富明亮的情景式商业空间。商务写字楼体形挺拔，立面简洁明快，顶部挑板的形象处理隐喻"飞翔"，与天翔公司的形象寓意不谋而合。

The tower building contains a 23-storey office building and 4-storey podium structure with the first basement level for lease and the second level for car parking. Considering that the development is occupied by a 5-storey commercial space including the first basement level, the architectural concept design proposed an inner pedestrian street to settle down the disadvantages of the depth of the shop front and introduce a bright and dramatic business space. The appearance of the office building is designed in a simple and straightforward form with the cantilevered panels on the top level addressing the sense of flight which corresponds to the spirit of the company.

杭州汇丽绣花制衣综合楼

Huili Embroidery and Clothing Complex，Hangzhou

综合楼位于萧山经济开发区的汇丽绣花制衣厂区内，东西宽 59.6m，南北深 89.6m，形体方正简洁，建筑高 3 层，局部 4 层，集展示、洽谈、娱乐、办公于一体。底层南面 20m 进深、14.5m 高的大厅，三面均为玻璃幕墙，南北二侧金属材料装饰的列柱，加上透明观光电梯、新颖玻璃楼梯和翠绿挺拔毛竹点缀，使整个大厅显得高贵、现代又生机勃勃，充满时代的气息。二层 40m × 24m 的中庭，洒满阳光，在绿化陪衬下，亲切宜人，是人们休息、交往的空间，也可兼作时装表演的场所。

The complex contains 3 levels and one mezzanine level accommodating exhibition, meeting, entertainment and office functions. The ground floor to the south is dominated by a lobby of 20 meter long and 14.5 meter high featured powerful glass curtain walls on three sides, metal-clad colonnades erected on each side, together with an observatory lift, the innovative glass stair and the green bamboo that are carefully integrated into the interior with the surrounding bamboo trees which enhances the dramatic and magnificent experience of the lobby. The two-storey atrium of 40 meter long and 24 meter wide is designed as an inviting gathering space or sometimes as the fashion show function that is complemented by the landscaping elements.

建 设 单 位：杭州汇丽绣花制衣有限公司
功 能 用 途：办公、娱乐、展示
设计 / 竣工年份：2001 年 /2001 年
建 设 地 点：杭州萧山城北经济开发区
总 建 筑 面 积 ：11700m²
总建筑高度 / 层数：15.4m/3 层
结 构 形 式：框架结构

Client: Hangzhou Huili Embroidery and Clothing Co., Ltd.
Function: Office, Entertainment, Exhibition
Designed/Completed: 2001/2001
Location: Economic Development Zone, Xiaoshan (North), Hangzhou
Total Floor Area: 11,700m²
Total Height/Floor: 15.40m/ 3 Floors
Type of Structure: Framed Structure

杭州东方通信大厦

Eastern Communication Tower, Hangzhou

建 设 单 位：杭州东方通信有限公司
功 能 用 途：办公
设计 / 竣工年份：1991 年 /1996 年
建 设 地 点：杭州市文三路
总 建 筑 面 积 ：32000m²
总建筑高度 / 层数：97.6m/22 层
结 构 形 式：框剪结构

Client: Hangzhou Eastern Communication Co., Ltd.
Function: Office
Designed/Completed: 1991/1996
Location: Wensan Lu, Hangzhou
Total Floor Area: 32,000m²
Total Height/Floor: 97.60m/ 22 Floors
Type of Structure: Framed Shear Wall Structure

杭州东方通信大厦是一座集科技、培训、商贸、办公等为一体的现代化智能型综合大厦。大厦总体布局合理、功能分区明确，各出入口设置及流线分明，互不干扰。介于“点式”和“板式”之间的外形，两侧逐渐收进，形成曲折有致的平面和高低错落的体形。该大厦已获钱江杯优秀建筑设计铜奖，同时，该项目已入编《杭州新建筑》一书。

The development offers a mix-use intelligent complex including science, training, business and office components. All functions within the well-proportioned interior spaces are clearly defined. All entrances and circulations are properly outlined and separated from each other. Respectful of the immediate surrounding and the significance of its functions, the appearance of the complex is designed in a manner that it employs the prototype of the spot and bar building and is set back at both ends featuring a well-designed curved facade and scaled in height. The project was awarded the honor of Copper Prize of Excellent Architectural Design of Qianjiang Cup and was included in the book of Hangzhou New Buildings.

宁波市财税征收管理中心

Ningbo Municipal Taxation Management Center, Ningbo

建 设 单 位：宁波市财税局
功 能 用 途：办公
设计 / 竣工年份：1998 年 /1999 年
建 设 地 点：宁波市中山西路
总 建 筑 面 积 ：45800 m^2
总建筑高度 / 层数：99m/27 层
结 构 形 式：框剪结构

Client: Ningbo Municipal Taxation Bureau
Function: Office
Designed/Completed: 1998/1999
Location: Zhongshan Xi Lu, Ningbo
Total Floor Area: 45,800m^2
Total Height/Floor: 99m/ 27 Floors
Type of Structure: Framed Shear Wall Structure

宁波市财税征收管理中心位于宁波市中山西路。该建筑由 27 层主楼、11 层副楼、4 层裙房组成，主楼与副楼之间以连廊相接。在功能上主楼为办公，副楼为招待所，裙房一、二层为征税大厅，三、四层为餐厅和职工之家，地下2层为错层式汽车库和设备用房，各部分功能配置明确、流线清晰。

The complex is composed of three main functions: the 27-storey main building dominated by offices, an 11-storey auxiliary building accommodating the guesthouses and the 4-storey podium base. Podium structure houses taxation hall on first and second floor and cafeteria and the staff' s home on third to forth floor. Car parking and mechanic rooms are located on the two basement levels. All functions are clearly defined and circulation system is meticulously designed.

上海银峰大厦

In-fong Office Tower, Shanghai

建 设 单 位：上海银峰商城发展有限公司
功 能 用 途：办公
设计 / 竣工年份：1994 年 /2004 年
建 设 地 点：上海浦东张杨路商业区
总 建 筑 面 积 ：56230m^2
总建筑高度 / 层数：165m/41 层
结 构 形 式：框筒结构
合作设计单位: 加拿大 PPA 设计所

Client: Silver Peak Development （Shanghai） Co., Ltd.
Function: Office
Designed/Completed: 1994/2004
Location: Zhangyang Lu,Pudong,Shanghai
Total Floor Area: 56,230m^2
Total Height/Floor: 165m/ 41 Floors
Type of Structure: RC Framed Core-wall structure
In associate with Petroff Partnership Architects

上海银峰大厦是上海浦东张杨路商业区的一部分。在平面与空间设计中，以室内步行街形式强化了裙楼与周围商业建筑的空间联系与渗透。在交通组织中针对超高层建筑的特点，将客梯以四台为一组分高低区运行，使人流快速到达各楼层。在立面造型设计中，运用方圆、曲直对比，并以对比中的和谐共存求得丰富而生动的形体变化，整个大楼体现出既坚实又挺拔的不凡气质。

The complex contains office space, retail, food & beverage and entertainment facilities with two basement levels for M&E rooms and car parking area. The simple geometry of a square within a cylinder was used to generate the strong impact the designs have on the Pu-Dong skyline. The tower entrance plaza has been enclosed to create a “crystal pavilion” retail court.

宁波中农信国际商贸大楼

CADTIC International Plaza, Ningbo

建 设 单 位：宁波之信房地产公司
功 能 用 途：商场、办公
设计 / 竣工年份：1994 年 /1996 年
建 设 地 点：宁波中山路
总 建 筑 面 积 ：80000m²
总建筑高度 / 层数：138m/38 层
结 构 形 式：框剪结构

Client: Ningbo Zhixin Real Estate Corp.
Function: Retail, Office
Designed/Completed: 1994/1996
Location: Zhongshan Lu, Ningbo
Total Floor Area: 80,000m²
Total Height/Floor: 138m/ 38 Floors
Type of Structure: Framed Shear Wall Structure

宁波中农信国际商厦坐落在宁波中心商业区中山路与三江口交叉口转角广场。设计根据功能的要求做到了合理的分区，近 3 万 m² 面朝中山路的裙房为宁波华联大型商场，37 层高的主楼为办公商务用房，西北角和义路设有单独入口。平面设计分别采用直径为 36m、16m、5m 的圆形筒体和连贯的曲面与曲线组成。观光电梯直达顶部餐厅，顾客可直接观赏三江口和市中心繁华景观。直径 16m 的中筒内设有垂直交通管道和六部电梯，主楼的避难层设在 23 层，顶层可停直升机。整个大厦构思新颖、造型美观，构筑了活泼的商业建筑形象。

All functions within the complex are clearly defined. The podium base facing Zhongshan Lu is dominated by Ningbo Hualian Department Store with an overall floor area of 30,000 square meters. The 37-storey main building accommodates the office building and is provided with a dedicated entrance to the northwest corner besides Heyi Lu. The appearance of the complex is designed in the circular core structure and curved profile. Observatory lift can directly reach to the restaurant on roof level where guests can enjoy a panoramic view towards San Jiang Kou and the downtown's prosperity. The core of 16 meter diameter is installed with vertical circulation system and 6 elevators. The refugee area is located on Level 23. Helicopter apron is served on the roof level. The architecture and planning of the complex are unique in the concepts and skillful in their execution recalling a lively urban ensemble in the center of the city.

温岭总商会大楼

General Chamber of Commerce, Wenling

建 设 单 位：温岭市工商联总会
功 能 用 途：写字楼、会议、展览、公寓
设计 / 竣工年份：2002 年 /2006 年
建 设 地 点：浙江省温岭市万昌路
总 建 筑 面 积 ：47000 m^2
总建筑高度 / 层数：99m/29 层
结 构 形 式：框剪结构

Client: General Chamber of Industry and Commerce, Wenling
Function: Office, Convention, Exhibition, Apartment
Designed/Completed: 2002/2006
Location: Wanchang Lu, Wenling, Zhejiang
Total Floor Area: 47,000m^2
Total Height/Floor: 99m/ 29 Floors
Type of Structure: Framed Shear Wall Structure

温岭总商会是集写字楼、公寓、餐饮、会议、展览等功能为一体的综合性建筑。大楼建筑平面形式充分考虑到地块形状、景观朝向等多方面要求，采用两个半圆相对的标准层平面及一个中心筒形式，一边为写字间，一边为公寓，是针对现阶段新类型建筑平面的一种尝试。

建筑造型简洁，外墙采用较大面积开窗和仿金属涂料墙面的组合，并设置建筑遮阳挑板，增强建筑的光影效果和节奏感，突出其商业特征。同时通过建筑体块的组合、曲线的应用和整体虚实、高低处理，形成不同体块特征的丰富建筑效果。

The complex, accommodating the office, apartment, restaurant, convention and exhibition hall, is designed to maximize the effective and efficient use of the site. Two semi-circular shaped structures are opposite located and dominated by office building and apartment building separately with a core structure between them. The appearance of the building is simple in its form and decorated with large operable windows and nature metal finishes. The cantilever balconies are a permanent and effective sunscreen. The articulation of the geometric patterns, the use of curvilinear and the design of solid and void space makes itself a new landmark in the city.

黄岩总商会大厦

General Chamber of Commerce, Huangyan

建 设 单 位：黄岩总商会
功 能 用 途：办公
设计 / 竣工年份：2002 年 /2006 年
建 设 地 点：台州市黄岩区城关
总 建 筑 面 积 ：40253.3m^2
总建筑高度 / 层数：88.8m/26 层
结 构 形 式：框剪结构

Client: General Chamber of Commerce,Huangyan
Function: Office
Designed/Completed: 2002/2006
Location: Cheng Guan Bei Lu, Huangyan
Total Floor Area: 40,253m^2
Total Height/Floor: 88.80m/ 26 Floors
Type of Structure: Framed Shear Wall Structure

本工程位于台州市黄岩区城关环城北路。大厦建筑形体上对称的构图及其四个转角上钻石型的办公空间的设计，让使用者充分享受到周边优美的人文环境，创造了现代景观办公空间，挺拔的建筑形体也成为环境中的至高点和统领环境要素的标志物与对景。

The overall site planning arrangement acknowledges the historic Chinese planning of space symmetrically organized. Four diamondshaped office structures are located at the top of the main building. This allows for the users to enjoy the immediate cultural and environment context and brings a sophisticated landscaped office space. The straightforward form of the complex dominates the view of the site as a landmark to its neighbor.

杭州市商业银行

Commercial Bank, Hangzhou

建设单位：杭州市商业银行
功能用途：办公、银行营业
设计/竣工年份：2003年/在建
建设地点：杭州市庆春路与建国路交界处
总建筑面积：67328m²
总建筑高度/层数：130m/31层
结构形式：框架核心筒结构
合作设计单位：澳大利亚HC建筑设计公司

Client: Hangzhou Commercial Bank
Function: Office,Bank Business
Designed/Completed: 2003/under constraction
Location: Intersection of Qingchun Lu & Jianguo Lu, Hangzhou
Total Floor Area: 67,328m²
Total Height/Floor: 130m/ 31 Floors
Type of Structure: Framed Core-wall Structure
In association with HC Consultant Pty. Ltd.

杭州市商业银行位于杭州庆春路与建国路交汇处的东北角，总用地面积为8547.36m²。主楼为银行办公用房及写字楼，裙楼为营业用房，其中营业大厅三层通高，气势恢宏，另设有餐饮、会议、多功能厅等配套用房及屋顶花园。地下3层设置金库、设备用房等，共配置泊车位300多个。大楼一层至九层采用冰蓄冷加电热蓄热为冷热源的集中式中央空调主机方式，十层以上主楼采用直接蒸发式变频多联空调系统，具有优良的综合节能效益。大楼采用5A智能化办公系统。

The development covers a total site area of 8,547.36 square meters. The main structure includes the offices for bank and for lease. Podium base is dominated by the business functions including the 3-storey high lobby, restaurant, meeting room, ball room and roof garden with 3 basement levels occupied by the coffers, the mechanical rooms and car parking for 300 cars. The central air-conditioning units are provided to the first to 9th floor completed with ice-storage air-conditioning system as cooling source and electric heat-storage air-conditioning units as heating source. Above the 10th floor is provided with direct evaporative inverter multi variable central air-conditioning unit. The project is featured effective and efficient energy-saving design and 5A building intelligent system.

海宁市行政中心

Haining Administrative Center, Haining

海宁市行政中心位于市区的南部，总用地面积23.5hm^2，是一组集市政广场、行政中心办公楼、会议接待中心、办证人才中心、档案馆、餐饮中心、轻印中心、后勤服务中心为一体的智能化办公建筑。建筑造型严格遵循总体设计的要求，既体现了政府建筑的庄重性，又表现出了现代感。该建筑2005年荣获浙江省优秀设计三等奖。

Located at the southern downtown, the development covers a total site area of 23.50 hectares and is designed as an intelligent building including the civic plaza, administrative office, meeting reception center, certificate approval center, human resource center, files and archives, restaurant, copy and printing center and back-of-house premise. The complex is designed for maximum efficiency to comply with the overall planning criteria that reflects the magnificent and the modern image of the government building. The project was awarded the honor of Third Prize of Excellent Design Zhejiang Province 2005.

建设单位：海宁市政府
功能用途：办公
设计/竣工年份：2001年/2003年
建设地点：海宁市
总建筑面积：71823m²
总建筑高度/层数：49.5m/12层
结构形式：框架、框剪结构

Client: Haining Municipal Government
Function: Office
Designed/Completed: 2001/ 2003
Location: Haining, Zhejiang
Total Floor Area: 71,823m²
Total Height/Floor: 49.50m / 12 Floors
Type of Structure: RC Framed & Framed Shear Wall Structure

宁波出入境检验检疫局综合实验楼

Mix-used Laboratory of Ningbo Entry-Exit Inspection and Quarantine,Ningbo

建 设 单 位：宁波出入境检验检疫局
功 能 用 途：实验楼
设计／竣工年份：2000 年 /2004 年
建 设 地 点：宁波市
总 建 筑 面 积 ：36035m^2
总建筑高度／层数：96.3m/23 层
结 构 形 式：框筒结构

Client: Ningbo Entry-Exit Inspection and Quarantine
Function: Laboratory
Designed/Completed: 2000/2004
Location: Ningbo, Zhejiang
Total Floor Area: 36, 035m^2
Total Height/Floor: 96.30m/23 Floors
Type of Structure: Framed Core-wall Structure

宁波出入境检验检疫综合实验楼是宁波市重点工程，位于中山西路与马园路交叉口，基地面积约12200m^2，该工程由主楼、副楼、连廊、地下停车库等几部分组成。主楼地下1层，地上23层，总高度为96.3m，副楼地下1层，地上4层，总高度18.9m。工程总建筑面积为36035m^2，综合实验楼一、二层为对外的报验大厅和报验资料室，三层为多功能厅、会议室，四层为电化教室、职工图书阅览室，五～十五层为办公室，十六～二十三层为各类实验室。副楼一层为综合用房，二层为职工餐厅，三～四层为办公室。主副楼之间设有连廊，便于工作联系。

Located at the intersection of Zhongshan Lu(West)and Mayuan Lu, the development covers a total slte area of 12,200 square meters including the main building, subsidiary building, link corridor and basement car park. The 96-meter-high main building consists of 23 floors aboveground and one basement level and the 19- meter- high subsidiary building with 4 floors above grade and one basement floor is connected to the main by link corridors.

The complex is dominated by declaration lobby and declaration files on the first and second floor. The third floor is occupied by ballroom and seminars. The forth floor includes the audio-visual classroom and library for staff. Office space is distributed on 5th to 15th floor. Laboratories of various kinds are located on 16th to 23rd floor. First floor in the subsidiary building is served by multi-used premise and second floor is occupied by staff canteen. From 3rd to forth floor contains the offices.

温州松台大厦

Songtai Tower, Wenzhou

建 设 单 位：温州联合房地产开发有限公司
功 能 用 途：商场、写字楼
设计 / 竣工年份：2001 年 /2004 年
建 设 地 点：浙江省温州市
总 建 筑 面 积 ：32000m²
总建筑高度 / 层数：88.8m /23 层
结 构 形 式：框剪结构

Client: Wenzhou United Real Estate Development Co., Ltd.
Function: Retail, Office
Designed/Completed: 2001/ 2004
Location: Xin He Lu, Wenzhou, Zhejiang
Total Floor Area: 32,000m²
Total Height/Floor: 88.80m /23 Floors
Type of Structure: Framed Shear Wall Structure

本工程位于温州市城区信河路松台广场南端，西北角为松台山，南临五星级酒店华侨饭店。

为满足尽可能多地安置回迁户的要求，设计了贯穿基地的一条内部步行街，实现了几乎占据整个基地的大裙房。步行街的设置减小了裙房的进深，使所有的商业店铺均为沿街，最大限度满足了安置的要求。主楼形体挺拔大方，注重虚实对比，北面观光电梯与楼顶观光茶室的标志性造型，成为松台广场南侧一道亮丽的风景。

Predominantly situated at the south end of Songtai Square and adjacent to Songtai Mountain to the north-west corner and the 5-star Overseas Hotel at the south, the proposed building has been designed with a pedestrian street through the east to the west and a podium base of monumental proportion that would settle down more dismantled family as possible. Pedestrian Street is designed to minimize the depth of thc podium structure that allows all shops facing the street level. The main building is simple and straightforward in its form and provided with prospective elevator and tea house on the top level which creates a dramatic new landmark on the city's skyline.

杭州威陵大厦

Weiling Building, Hangzhou

建 设 单 位： 杭州滨海房地产开发有限公司
功 能 用 途： 办公
设计 / 竣工年份： 2001 年 /2005 年
建 设 地 点： 杭州市滨江区
总 建 筑 面 积： 50080m²
总建筑高度 / 层数： 57～78m/13～22 层
结 构 形 式： 框架剪力墙结构

Client: Hangzhou Binhai Real Estate Development Co., Ltd.
Function: Office
Designed/Completed: 2001 / 2005
Location: Binjiang District, Hangzhou
Total Floor Area: 50,080m²
Total Height/Floor: 57m～78m/ 13～22 Floors
Type of Structure: Framed Shear Wall Structure

本工程位于杭州滨江区中心路北侧，靠近滨江区区政府，分为 1 号、2 号两个地块，其中 1 号地块建筑用地面积为 5525m²，为滨海大厦用地，2 号地块建筑面积 5828m²，为威陵集团办公用地，地块北侧为住宅区，东侧为财税局办公楼。根据业主要求及结合规划条件，1 号、2 号地块统一开发建设，两地块合并设两层地下室，共设 286 辆机动车停车位，供两地块共同使用，1 号地块上设 22 层点式办公楼，共计 18378m²，2 号地块上设 13 层板式办公楼，共计 17802m²。整个建筑群设计风格力求清新自然、造型简洁明快，整个工程成为一个充满现代气息的建筑群。

The project is divided into two lots and developed under the overall master planning design. Lot One covers an area of 5,525 square meters dominated by the 22-storey Binghai Tower of 18,378 square meters. Lot Two occupies an area of 5,828 square meters located by the 13-storey bar-shaped Weiling Group Office Building of 17,802 square meters. Two basement car parking area is provided for 286 cars and shared by the two lots. The architectural design strives to exhibit its sophisticated simplicity and creates a comprehensive state-of-the art ensemble.

苏州新区科技创业园创业大厦

Hi-tech Incubator Center Incubator Building • Suzhou New District,Suzhou

建设单位：苏州新区科技创业园
功能用途：办公
设计 / 竣工年份：2002 年 /2004 年
建设地点：苏州市
总建筑面积：32000m²
总建筑高度 / 层数：23.9m/6 层
结构形式：框架结构

Client: Hi-tech Incubator Center Suzhou New District
Function: Office
Designed/Completed: 2002/2004
Location: Suzhou
Total Floor Area: 32,000m²
Total Height/Floor: 23.90 m/ 6 Floors
Type of Structure: Framed Structure

苏州新区科技创业园是以创业中心为核心，创业大厦是整个创业园的核心，里面设置了众多的公共服务功能：展厅、餐厅、信息中心、管理中心、商务中心、会议室、公共实验室等。创业大厦是建筑与自然密切结合的有机体，在城市道路转角处架空退让。建筑立面上的大片通透玻璃与遮阳板的结合体现了对周边良好自然环境的尊重，绿化自然地渗透进入室内。总体的布局遵从创业园整体的功能要求，建筑沿竹园路呈舒展优美的曲线，丰富了街景，并与周围环境有机地融合在一起。

As a hub in Hi-tech Incubator Center Suzhou New District, the development contains public service functions of various kinds such as exhibition, restaurant, information center, administrative center, business center, seminar and laboratory, etc. The concept design for the ensemble integrates the architecture with the nature. The facades are a combination of transparent glass and sunscreens that is respectful of the natural landscape. It allows the green elements permeating the building. The architectural composition arrangement refers to the functional requirements, simultaneously melds into the immediate context. The soft undulating curve of the building along Bamboo Garden Road is designed that animates the street level and incorporates into the surroundings.

萧山高运集团大厦

Gaoyun Group Building, Xiaoshan

建 设 单 位：浙江高运建设集团
功 能 用 途：写字楼
设计 / 竣工年份：2005 年 / 在建
建 设 地 点：萧山区金成路、育才路
总 建 筑 面 积 ：22200m²
总建筑高度 / 层数：81.60m/22 层
结 构 形 式：框剪结构

Client: Zhejiang Gaoyun Construction Group
Function: Office
Designed/Completed: 2005/under construction
Location: Jincheng Lu,Yucai Lu,Xiaoshan, Zhejiang
Total Floor Area: 22,200m²
Total Height/Floor: 81.60 m / 22 Floors
Type of Structure: Framed Shear Wall Structure

本工程位于萧山区金成路与育才路交叉口西南角，交通便利，为萧山区北干派出所与浙江高运集团联建项目。裙房一层至六层为北干派出所业务用房，主楼七层以上为高运集团办公用房。设计赋予了建筑物东、北两个主立面两种性格。东立面为北干派出所主入口，建筑为强烈、虚实对比的手法，创造出庄重大气的建筑表情。北立面直接面对城市主干道金成路，设计采用挺拔的铝合金竖线条结合玻璃幕墙的处理手法，造型简洁。

The proposed building enjoys an accessible traffic to it and includes 6-storey podium base for Xiaoshan Beigan Police Station and 7-storey main building for Gaoyun Group. The design suggests two kinds of facades to the east and north of the building. As the main entrance to Beigan Police Station, the facade of the building at the east is formal and imposing, yet at the north to the main boulevard the building has facade of aluminum and glass that reaches upward in a simple and straightforward progression.

杭州广利大厦

Guangli Mansion, Hangzhou

建设单位：广利集团
功能用途：办公
设计/竣工年份：2002年/2005年
建设地点：杭州庆春路与中河路交叉口处
总建筑面积：62968m²
总建筑高度/层数：99.6m/26层
结构形式：框架－核心筒结构
合作设计单位：上海现代建筑设计集团

Client: Hangzhou Guangli Group
Function: Office
Designed/Completed: 2002/ 2005
Location: Intersection of Qingchun Lu & Zhonghe Lu, Hangzhou
Total Floor Area: 62,968m²
Total Height/Floor: 99.60m / 26 Floors
Type of Structure: Framed Core-wall Structure
In association with Shanghai Xiandai Architectural Design （Group） Co., Ltd.

广利大厦位于杭州庆春路金融街与杭城南北交通大动脉中河路交汇处，总用地面积8820m²。大厦是现代化高档纯商务写字楼。底层不设商铺，且充分配备服务于商务人士的商务中心、休闲吧等设施。裙楼屋顶有屋顶花园，主楼屋顶有观光层，创造出写字楼的休闲价值。地下3层，共配置泊车位300余个。大厦主要采用Hitachi分层式中央空调，分层计量，采用5A智能化办公系统。建筑通体采用双层中空钢化玻璃幕墙。

This modern complex covers a total site area of 8,820 square meters including offices, the business center, lounge bar and roof garden on the podium structure. An observation level is occupied on the top of the main building. Three basement levels are provided for 300 cars. The development is equipped with the state-of-the-art Hitachi district VRV air-conditioning system and metering by floors. It is also completed with 5A Building Intelligent System. The exterior is enveloped with tempered double glazed curtain wall.

杭州西子发展大厦

Xizi Development Mansion, Hangzhou

杭州西子发展大厦位于杭州学院路，是一栋集展示、科研办公、生活配套为一体的综合楼。大楼建成后，将成为德力西集团有限公司的总部大楼。

建筑由一幢90m高的办公楼与75m高的配套楼构成，两楼的错位布局形成前后两广场，使内外流线有合理的分离，建筑日照视野亦因此得以双赢。在满足功能要求的前提下，设计着重塑造了建筑形象的庄重大方，简洁挺拔的竖向线条，稳重大方的黑色石材外墙，体现了业主雄厚的实力和稳健踏实的企业文化。

The proposed building is composed of exhibition, office for scientific research and supporting functions for living and would serve as headquarter of DELIXI Group Co., Ltd. It contains a 90-meter-high main building and a 75-meter-high auxiliary building which are arranged in staggered configuration, creating two squares at its back and front that allows for a clearly defined circulation, well designed nature light and visual experience. On the premise of the function requirement, the architecture of the building strives to reflect its formal and imposing form complemented with simple straightforward lines and the modest black stone cladding that embodies the strength and spirit of the corporation.

建设单位：杭州西子科技实业有限公司
功能用途：写字楼
设计/竣工年份：2004年/未建
建设地点：杭州学院路
总建筑面积：60000m²
总建筑高度/层数：90m/25层
结构形式：框筒结构

Client: Hangzhou Xizi Scientific & Technologic Industrial Corporation Limited
Function: Office
Designed/Completed: 2004/unbuilt
Location: Xueyuan Lu, Hangzhou
Total Floor Area: 60,000m²
Total Height/Floor: 90m / 25 Floors
Type of Structure: Framed Core-wall Structure

绍兴永利大厦

Yongli Tower, Shaoxing

建 设 单 位：永利房地产有限公司
功 能 用 途：写字楼
设计 / 竣工年份：2003 年 /2005 年
建 设 地 点：绍兴柯桥大道以西、钱陶公路以南
总 建 筑 面 积 ：44492m²
总建筑高度 / 层数：99.4m/26 层
结 构 形 式：框架核心筒结构

Client: Yongli Real Estate Development Co., Ltd.
Function: Office
Designed/Completed: 2003/2005
Location: Shaoxing, Zhejiang
Total Floor Area: 44, 492m²
Total Height/Floor: 99.40m/26 Floors
Type of Structure: RC Framed Core-wall Structure

永利大厦位于绍兴柯桥CBD(中央商务区)中心入口转角处，它的建成将成为绍兴地区的标志性建筑之一。平面布置根据用地状况成不规则弧形与矩形相结合，建筑的界面与城市环境取得了有机的联系，建筑构图形式新颖、活泼，功能与环境结合，并融办公、观景和商务等诸多服务功能及规划为一体的大厦。

大厦标准层平面 1300m²，核心筒中布置 6 台高速电梯，地下室建筑面积 5415m²。

The project is located at the corner of CBD entry of Keqiao Shaoxing. Incorporating the comprehensive program of office, commercial and business functions, the tower creates a dramatic landmark on the city's skyline. The combination of irregular circular and rectangular form melds the site and architecture into a single, comprehensive expression that organically integrates the building with the urban environment. Its contemporary form is an ideal complement to the immediate context. Typical floor area is estimated to be 1,300 square meters. Six high-speed elevators are provided in the core structure. Basement level occupies a floor area of 5,415 square meters.

浙江华越生产研发大楼

Zhejiang Huayue Software Production, Research & Development Center, Hangzhou

建设单位：浙江华越控制软件有限公司
功能用途：生产、办公
设计/竣工年份：2005/在建
建设地点：杭州文二路
总建筑面积：24681m²
总建筑高度/层数：87.7m/25层
结构形式：框架核心筒结构

Client: Zhejiang Huayue Controlling Software Development Co., Ltd.
Function: Production, Office
Designed/Completed: 2005 /under construction
Location: Wen' er Lu, Hangzhou
Total Floor Area: 24,681m²
Total Height/Floor: 87.70m/ 25 Floors
Type of Structure: Framed Core-wall Structure

本工程是位于杭州老城区内的高新企业的新建生产、办公建筑。考虑到周边紧密相连的公共及住宅建筑，建筑师建议业主放弃了会产生严重光污染的玻璃幕墙的外立面设计，采用了同样简洁且可以反映企业现代形象的立面处理。考虑到对北侧住宅日照的影响，建筑形体由底层向上部三次退台，从而也造就了多变的建筑形象。

Located in the downtown of the old city, the concept design for the development considered the existing public and residential buildings that are closely around it. It rejected the use of the curtain wall that might create optical pollution to the immediate buildings. The architectural design also suggests a sophisticated simplicity to the facades of the building that enhances the prestige and spirit of the corporation. Considering the impact on the day light to the residences at the north, the building steps in from the ground floor three times to achieve spatial diversity and formal richness.

义乌福田大厦

Futian Mansion, Yiwu

建 设 单 位：浙江伊厦房地产开发有限公司
功 能 用 途：商业、办公
设计 / 竣工年份：2004 年 /2006 年
建 设 地 点：浙江省义乌市
总 建 筑 面 积：46870m^2
总建筑高度 / 层数：99.9m/ 地上 27 层
结 构 形 式：框架结构

Client: Zhejiang Yixia Real Estate Development Co., Ltd.
Function: Retail, Office
Designed/Completed: 2004/2006
Location: Yiwu, Zhejiang
Total Floor Area: 46,870m^2
Total Height/Floor: 99.90m/ 27 Floors
Type of Structure: Framed Structure

福田大厦位于义乌市福田新区内，是国际商贸城内的重要建筑。建筑地上27层，地下2层，主要功能为商业和办公。高层主体位于用地的北侧，裙房则沿西侧道路展开，呈长方形布局。建筑体型简洁，线条流畅。

Strategically situated in Yiwu International Trade Town, the project is composed of the commercial and office functions with 27 floors aboveground and 2 floors underground. The square high-rise main building is located at the south and the simple rectangular podium base is extended to the west.

LIGHT

浙江华越软件发展中心

Zhejiang Huayue Software Development Center, Hangzhou

建 设 单 位：浙江华越控制软件有限公司
功 能 用 途：生产、办公
设计 / 竣工年份：2004 年 / 在建
建 设 地 点：杭州市滨江区江南大道南侧
总 建 筑 面 积 ：33265m²
总建筑高度 / 层数：100m/26 层
结 构 形 式：框架剪力墙结构、框架核心筒结构

Client: Zhejiang Huayue Controlling Software Development Co., Ltd.
Function: Production, Office
Designed/Completed: 2004/under construction
Location: Hi-tech Development Zone, Binjiang District, Hangzhou
Total Floor Area: 33,265m²
Total Height/Floor: 100m/ 26 Floors
Type of Structure: Framed Shear Wall Structure, Framed Core-wall Structure

工程位于杭州市滨江高新区江南大道南侧。本项目地处以高新产业为代表的新兴城市区块，建筑形式即以数字化时代崇尚的简洁为基调，为业主塑造了作为控制软件公司所具有的诚实、可靠、稳重的公司形象。在此基础上建筑采用新材料、新技术，表现出现代且不失高雅的建筑风格。

Located in the new developed district of the city, the architecture of the building is designed to convey the sense of digital times. It is simple in its form but reveals the honest, reliable and prudent aspect of this hi-tech company. Additionally it strives to reflect new technology in building systems and materials that creates a state-of-the art and graceful identity for the location.

北京正阳大厦

G-Box, Beijing

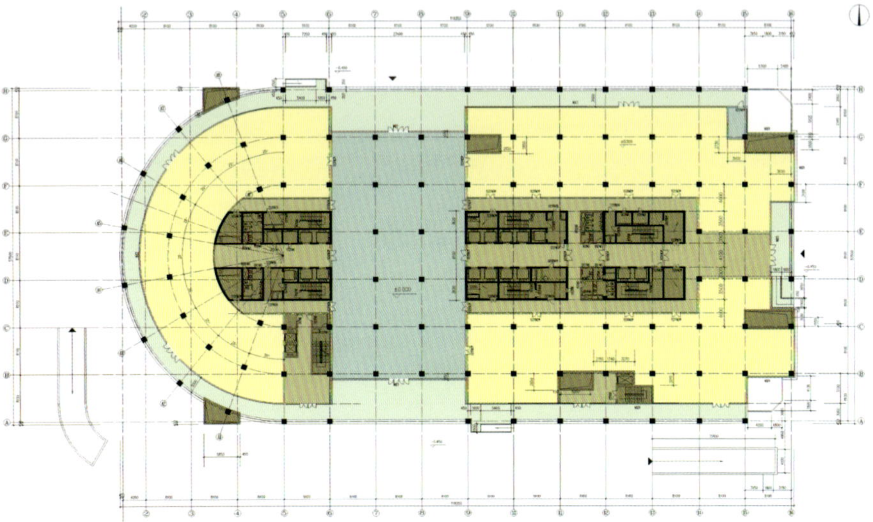

建设单位：北京美光房地产开发有限公司
功能用途：高科技写字楼
设计/竣工年份：2001年/2006年
建设地点：北京市
总建筑面积：107670m²
总建筑高度/层数：60m/16层
结构形式：框架核心筒结构

Client: Beijing Meiguang Real Estate Development Co., Ltd.
Function: Office
Designed/Completed: 2001/ 2006
Location: Beijing
Total Floor Area: 107,670m²
Total Height/Floor: 60 m / 16 Floors
Type of Structure: Framed Core-wall Structure

本工程位于北京市北二环内，紧邻雍和宫。建筑包括地下3层，地面16层，功能为北京歌华集团高科技产业写字楼，已列入北京2008年奥运会十大新文化建筑，主要平面功能为：地下二、三层为停车库，地下一层为设备用房，地面一、二层为营业用房，三层以上为网络机房及写字楼。建筑总高度为60m。立面造型简洁大气，体块虚实变化强烈，形成较强的气势，通过不同材料的组合，在和谐统一中又不失丰富的细部刻画。

Located at the North II Ring Road near Yong He Gong, the development serves as the hi-tech office building for Beijing Gehua Group with 16 floors aboveground and 3 floors underground. It consists of business functions on the first and second floor. The computer room and offices are above the third floor. Mechanical rooms are provided on the basement level 1. The second and third floor below grade are occupied by the car parking. The building is simple in its form, rich and textured in its material and detail and incorporative of modern technology. The architectural design aims to reflect the magnificence and unification of the building.

上海普陀区人民政府

Putuo District People's Government, Shanghai

规划中最重要的主轴线呈斜45°连接交叉口、圆形广场和区政府楼，引导主要步行人流进入圆形广场。整个规划建筑群相互围合形成向心力极强的圆形空间，并以此向四周放射布置工商、后勤、公安、检察院等办公楼。整个建筑群的造型统一中有变化。洁白的色彩、统一的元素构成一种协调的美，而每个单体的不同造型又各具特色。

The main axis is master planned at an angle of 45 degree to connect the intersection, the circular square and the government building and to direct the pedestrian circulation to the circular square. Around the square is dominated by the industrial commercial building, service functions, public security and procuratorate office building. Each function is specifically designed, yet there is a unifying manner of design between them.

建设单位：上海市普陀区武宁地块开发办公室
功能用途：办公
设计/竣工年份：1998年/2000年
建设地点：上海市普陀区大渡河路、铜川路
总建筑面积：58623m²
总建筑高度/层数：26.8m/7层
结构形式：框架结构

Client: Wuning Development Office of Putuo District, Shanghai
Function: Office
Designed/Completed: 1998/2000
Location: Daduhe Lu, Tongchuan Lu,Putuo District,Shanghai
Total Floor Area: 58,623m²
Total Height/Floor: 26.80m/ 7 Floors
Type of Structure: Framed Structure

绍兴市中级人民法院审判办公大楼

Courtroom & Office Building of Shaoxing Intermediate People' s Court,Shaoxing

建设单位：绍兴市中级人民法院
功能用途：审判、办公
设计/竣工年份：2002年/2003年
建设地点：绍兴市和畅堂
总建筑面积：38257m²
总建筑高度/层数：40m/10层
结构形式：框架结构

Client: Shaoxing Intermediate People' s Court
Function: Courtroom & Office Building
Designed/Completed: 2002/2003
Location: He Chang Tang, Shaoxing
Total Floor Area: 38,257m²
Total Height/Floor: 40m /10 Floors
Type of Structure: Framed Structure

绍兴市中级人民法院位于绍兴市和畅堂与环城西路交叉口。建筑由审判A楼、办公B楼、法官培训中心三个主体楼形成一组建筑群，楼内连廊相接，院内道路畅通，绿化优美，入口分明，体现法院的府院特色。立面造型上采用黑白色调和体块穿插组合的手法。同时将法文化理念抽象为京剧中包公脸谱的黑面造型，应用到主体审判楼的立面体块中，将中国古建筑坡屋顶的飞檐抽象为一条曲线的女儿墙天际线，意图体现坡屋顶的一个动态和趋势，并与周边历史地理环境呼应。

The development is composed of courtroom, office building and training center for judger which are connected by corridors within it. Vehicular and pedestrian circulation has been master planned to be direct, simple and easily comprehended. The facade of the courtroom is finished with black and white tones complemented by a combination of geometric patterns evoking the facial makeup of Bao Zheng in Peking Opera, an upright official known for safeguarding the dignity law in Song Dynasty.

湖州市中级人民法院审判办公大楼

Courtroom & Office Building of Huzhou Intermediate People' s Court,Huzhou

建 设 单 位：湖州市中级人民法院
功 能 用 途：审判、办公
设计 / 竣工年份：2004 年 /2006 年
建 设 地 点：湖州市经三路以西，经二路以东，纬二路以北地块
总 建 筑 面 积：23510.70m²
总建筑高度 / 层数：49.50m/12 层
结 构 形 式：框剪结构

Client: Huzhou Intermediate People' s Court
Function: Courtroom & Office Building
Designed/Completed: 2004/2006
Location: Jingsan Lu on the West, Jing' er Lu on the East, Wei' er Lu on the North
Total Floor Area: 23510.70 m²
Total Height/Floor: 49.50m/12 Floors
Type of Structure: Framed Shear Wall Structure

本工程基地依山傍水，自然环境优美，又处于未来城市的经济、文化、行政中心地区，设计的重点放在创造一个与自然环境共生的建筑群体。

法院的审判楼敦实厚重，富雕塑感，办公楼是整个建筑群的至高点，统领全局。公众出入口设在南侧，庄严大气。人流从扇形广场，顺大台阶而上，进入10m通高的候审大厅内，阳光透过正面和顶部的玻璃洒在大厅中，象征了法律的光明和正义，同时也代表了人民法院司法办案的透明性与公开性。正入口两侧立面处理的灵感源于古代雕刻在石碑上的法典，一个个字符均匀地刻在石碑上形成最初法律的依据，而本设计立面上均匀小方窗在排列上模仿了字符在石碑上的排列，充满韵律感。

Due to its splendid location, the design for the development aims to provide an ensemble to be integrated into the nature. The courtroom appears as a representative sculpture. The office building is designed as the highest structure dominating the magnificent view of the site. As soon as one enters the site from the great public entrance at the south, its grandeur and monumentality is felt. Once ascending a monumental stair through the fan-shaped square, a 10-meter-high lobby for awaiting trial is provided. Nature light is introduced from the windows at the facades and the clerestories at the top which symbolizes the brightness and justice of the law, meanwhile represents the openness and publicity of judicial procedure. Facades at the two sides of the main entrance portal are carefully designed evolved from the code inscribed on the stone tablet at the ancient time that means the basis of the initial law. Small square windows are well-proportionally arranged on the facades imitating the arrangement of the letters of the code engraved on the tablets, conveying a sense of rhyme.

杭州嘉华国际商务中心

Jiahua International Business Center，Hangzhou

建 设 单 位：浙江嘉华置业投资有限公司
功 能 用 途：高档写字楼
设计/竣工年份：2001 年 /2003 年
建 设 地 点：杭州市杭大路、西溪路
总 建 筑 面 积：51900m²
总建筑高度/层数：49.82m/ 14 层
结 构 形 式：钢－混凝土结构

Client: Hangzhou Jiahua Real Estate Investment Co., Ltd.
Function: High-grade Office Building
Designed/Completed: 2001/2003
Location: Hangda Lu/Xixi Lu, Hangzhou
Total Floor Area: 51, 900m²
Total Height/Floor: 49.82m / 14 Floors
Type of Structure: Reinforced Concrete Structure

杭州嘉华国际商务中心为高档纯商务写字楼，位于杭州 CBD 黄龙区域内。建筑一层为展示厅，二层为共享商务休闲空间，三至十四层为写字楼，地下室配置员工餐厅和汽车库。建筑主立面以柔美、细腻的水平线条强化曲线形的建筑造型特征，着意体现大气、高贵、灵秀的风格，以诠释杭州城市特征。建筑设计注重人性化、自然化的内部空间与景观设计，沿杭大路设计参与性休闲绿色广场，在建筑二层设计 1200m² 的呼吸式阳光绿色中庭，在东立面七至十一层局部内凹形成空中花园，在屋顶平台设计与宝石山脉相呼应的景观花园。建筑采用钢－混凝土混合结构，由圆钢管混凝土柱－焊接工字形梁框架、混凝土剪力墙组成整体结构，楼板采用压型钢板上现浇混凝土的组合楼板，在带来可观经济效益的同时，充分展现建筑师要求的结构美。该项目获 2005 年度浙江省建设工程钱江杯二等奖。

Located at the CBD Dragon Zone, this high-grade mix-used development consists of exhibition hall on the first floor, business entertainment facilities on the second floor, office building on the third to 14th floor, cafeteria for the staff and car parking on the basement level. The facades are finished with soft horizontal bands to reinforce its curved shape that exhibits its graceful magnificent style and reinterprets the characteristics of Hangzhou city. The architectural composition focus on the human-oriented and natural environment. A landscaping plaza along Hang da Lu is designed and complemented with the 1,200-square-meter planted atrium lobby on the second floor, the concaved hanging garden on the 7th to 11th east facades as well as the roof garden that corresponds to Jade Mountain miles away. The complex is designed as the reinforced concrete structure integrated with round steel tube concrete column, welded H-beam frame structure and concrete shear wall that is confirmed maximum cost-effectiveness but also reflects the visual aesthetics required by the architects. The project was awarded the honor of Qianjiang Cup Zhejiang 2005.

萧山商会大厦

Chamber of Commerce, Xiaoshan

建 设 单 位：萧山总商会
功 能 用 途：办公、会议
设计 / 竣工年份：2004 年 / 在建
建 设 地 点：杭州市萧山金城路
总 建 筑 面 积：113748m²
总建筑高度 / 层数：100m/28 层
结 构 形 式：框筒结构

Client: General Chamber of Commerce, Xiaoshan
Function: Office, Convention
Designed/Completed: 2004/under construction
Location: Jingcheng Lu, Xiaoshan, Hangzhou
Total Floor Area: 113,748m²
Total Height/Floor: 100m / 28 Floors
Type of Structure: Framed Core-wall Structure

萧山商会大厦位于杭州市萧山区金成路以北的中心商务区，基地南临金城路，北靠大浦河。其主要服务对象为萧山总商会下辖的37家会员企业和总商会总部。使用性质为高档写字楼。大厦主体为两幢对称的塔式100m高层建筑，底层4层为裙房。建筑采用大空间设计，主要为展示、会议等功能。五层以上为正方形平面的标准层，主要为办公功能。大厦整体体形匀称、稳重，立面造型设计大方新颖。凸出的竖向冷色铝材饰面檩条与同样是冷色的玻璃窗和窗下铝板饰面共同形成的立面格局贯通到顶。塔楼的四个角部处理成折线，外贴白色花岗岩饰材，与顶层高耸的女儿墙相互连贯，为主立面勾勒出一个清晰的白色底框。

The proposed project is located in CBD of the city and provides service to the 37 members and the headquarter of the Chamber of Commerce Xiaoshan. The symmetrically organized twin towers includes offices with 4-storey podium base containing the exhibition and convention hall. Above the fifth floor is designed as the square-shaped standard floor. The architecture of the building strives to enhance the prestige and spirit of the owner through the well-proportioned interior space, the formal and imposing facades and the rich and textured materials. The 28-storey tower has facades of aluminum and glass that reach upward in a straightforward geometric progression. Four corners of the skyscraper is treated with folded lines in its detail and clad in white-tinted granite stone up to the parapet on the top that outlines the contour of a clear white frame on its facades.

宾馆饭店建筑
Hotel Building

杭州饭店

Hangzhou Hotel, Hangzhou

杭州饭店位于杭州市北山路岳王庙东侧。主体建筑为江南传统的歇山屋顶，两侧平屋顶采用传统的局部起翘屋檐，体形高低错落。南面主入口柱廊檐下有弧形车道和宽大踏步进入大厅。小会堂的主要功能是开会及演出。由于在设计指导思想上始终把处理好“环境、空间、建筑”三者的关系作为建筑创作构思的精髓，饭店建成后不仅与周围湖光山色十分和谐，而且与近旁的岳王庙古建筑群组成了有机的整体，充分体现了江南传统的建筑风格。

Hangzhou Hotel is situated at the southern shore of the scenic Westlake and bounded by Yue Temple to its west. Respectful of the indigenous vernacular aspects , the top of the main building is covered with traditional Chinese hip and gable roof and the plat roofs at each side are to be partial titled eaves. The composition is well proportioned in its scale and staggered in height. A circular driveway and spacious stairs under the eave supported by prostyle brings the guests from the main entrance to the lobby. Convention hall hosts the meeting and the performance. The goal of its design is to meld the site and architecture into a single, comprehensive expression and to incorporate the immediate context around Yue Temple into the development.

建 设 单 位：浙江省旅游局
功 能 用 途：宾馆、会堂
设计 / 竣工年份：1956 年 /1959 年
建 设 地 点：杭州市北山路岳王庙东侧
总 建 筑 面 积 ：29500m²
总建筑高度 / 层数：20m/6 层
结 构 形 式：砖混、框架结构

Client: Zhejiang Provincial Tourism Bureau
Function: Hotel
Designed/Completed: 1956/1959
Location: Yue Temple（East）, Beishan Lu, Hangzhou
Total Floor Area: 29,500m²
Total Height/Floor: 20m/ 6 Floors
Type of Structure: Brick Concrete Structure & Framed Structure

浙江普陀山大酒店

Mountain Putuo Hotel, Zhoushan, Zhejiang

酒店地处浙江舟山国家级风景区普陀山岛上，是一座按四星级建造的约有240间标准客房的旅游宾馆，宾馆内设施较齐全，有餐饮、会议、娱乐、桑拿、健身等设施。

建筑造型大胆吸取传统江南民居形式，并加以革新创造，融汇了现代手法，颇有新意，整个造型采用坡顶，高低错落有致，与环境协调。建筑平面布局紧凑，功能合理，是传统建筑形式运用于现代建筑的一次成功的探索与尝试。

The sophisticated 4-star hotel is located on the scenic Mountain Putuo Island accommodating 240 typical guest rooms and housing the food& beverage, convention, entertainment, sauna and health club, etc. The form of the hotel is to exude the spirit of indigenous residence and one of modern interpretation of the traditional Chinese architecture. It is designed with pitched roof and staggered height in order to blend into the immediate surroundings. All functions within are meticulously defined. Great effort has been made to create both a modern functional building as well as one of the obvious indigenous aspect.

建 设 单 位：普陀山大酒店筹建处
功 能 用 途：旅游宾馆
设计 / 竣工年份：1995 年 /1998 年
建 设 地 点：浙江省舟山市普陀区
总 建 筑 面 积：12000m²
总建筑高度 / 层数：12m/4 层
结 构 形 式：砖混结构

Client: Mountain Putuo Hotel Task Force
Function: Hotel
Designed/Completed: 1995/1998
Location: Putuo District, Zhoushan, Zhejiang
Total Floor Area: 12,000m²
Total Height/Floor: 12m/ 4 Floors
Type of Structure: Brick & Concrete Structure

杭州凯悦大酒店

The Hyatt Regency Hangzhou Hotel, Hangzhou

建 设 单 位：杭州华浦实业有限公司
功 能 用 途：五星级旅游饭店及公寓
设计／竣工年份：1996 年 /2005 年
建 设 地 点：杭州湖滨路
总 建 筑 面 积 ：120000m^2
总建筑高度／层数：31.6m/ 8 层
结 构 形 式：框架结构
合作设计单位：美国波特曼建筑设计事务所

Client: Hangzhou Huapu Industrial Co., Ltd.
Function: 5-Star Hotel & Apartments
Designed/Completed: 1996/ 2005
Location: Hubin Lu, Hangzhou
Total Floor Area: 120,000m^2
Total Height/Floor: 31.60m/ 8 Floors
Type of Structure: RC Framed Structure
In association with John Portman & Associates

杭州凯悦大酒店选址在杭州市区湖滨路步行街区，为沿湖重点景观地带。它是集五星级酒店、酒店式公寓和大型商场于一体的综合性建筑。设计前期由美国波特曼建筑设计事务所承担。由于杭州西湖风景区景观保护的限高要求，建筑采用钢筋混凝土预应力框架结构，以争取在有限总高度内实现一定的规模。总体布局沿西湖为半环形酒店区，面向西湖的前广场安排环形车道和游泳池，沿市区仁和路的东区为商场，上层是公寓，建筑上部采用阶梯式布局，并安排了大面积屋顶景观平台，使酒店和部分公寓可以眺望西湖景色。

酒店的餐饮、休闲等功能配套完善，并在地下室设大型多功能宴会厅和大型娱乐设施。地下室中除安排设备用房，还有两层汽车库和完善的物流和服务系统。

The Hyatt Regency Hangzhou Hotel is located on the eastern shore of the scenic Westlake. The complex contains three main components: luxury hotel, service apartment and department store. Respectful the planning requirements for the site by limiting the height of the buildings and assuring that the scale complements the neighboring buildings, the structure framing for the project is designed as reinforced concrete structure with post tensioning reinforcing to minimize structural framing depth and provided for maximum ceiling height.

The crescent-shaped plan of the hotel embraces the lake with open arms and maximizes the magnificent views for the hotel guests. Beyond the hotel, the apartments and a dynamic multi-level retail complex are provided.

The complex steps back to accommodate the grand roof terrace from which to enjoy the scenic views and beauty of Westlake.The basement levels of the hotel accommodate hotel offices, back-of-house operations and mechanical and electrical rooms and parking for bicycles and cars in addition to the ball and meeting room.

萧山太虚湖假日酒店

Taixuhu Holiday Inn Xiaoshan, Hangzhou

建 设 单 位：太虚湖假日酒店有限公司
功 能 用 途：酒店
设计 / 竣工年份：2005 年 / 在建
建 设 地 点：萧山区义桥镇东方文化园
总 建 筑 面 积：36124m²
总建筑高度 / 层数：28m/7 层
结 构 形 式：框架结构

Client: Taixuhu Holiday Inn Co., Ltd.
Function: Hotel
Designed/Completed: 2005/ under construction
Location: Oriental Cultural Park, Yiqiao County, Xiaoshan
Total Floor Area: 36,124m²
Total Height/Floor: 28m/ 7 Floors
Type of Structure: Framed Structure

该酒店为扩建项目，位于东方文化园内，基地西南侧为原有宾馆，东、西两侧为疗养院和康乐中心。新建酒店客房总数为 300 个自然间，是集餐饮、会议、娱乐于一体的五星级度假酒店。

方案采用扇形裙房与“L”形主楼相围合构成宾馆主体，建筑体形舒展大气，主楼采用平坡结合的形式，东南端作逐层退台与周边建筑与山体和谐共处。裙房采用平屋面，外墙干挂花岗石板，强烈的虚实对比，简洁流畅的形象，体现了高星级酒店典雅高贵的特性。主楼西面与南面客房均采用落地窗及观景阳台，色调以灰瓦白墙为主，与周边建筑与山体取得谐调。

Located in Hangzhou Oriental Cultural Park and bounded by the existing hotel at its southwestern side, this expanded hotel is to be designed with 300 room modules and 5-star grade. The crescent-shaped podium structure and the L-shaped main building are integrated into the main structure of the hotel. The building is elegant and magnificent in its form with the main building to be covered with flat and pitched roof structure and step in at the east and south that relates to the immediate buildings and surrounding environment, yet the podium base to be topped with flat roof and clad with the granite stone on exterior walls. The contrast of solid and void coupled with the simple form of the building enhances the graceful and elegant identity of this high quality product. The rooms on the west and south side enjoy French windows and balconies overlooking the scenic views. The hotel exterior cladding is white-colored wall topped with gray tiles to blend into the immediate buildings and the hilly terrain of the site.

余杭大厦

Yuhang Building, Hangzhou

建 设 单 位：余杭区房地开发总公司
功 能 用 途：宾馆、会议、餐饮
设计 / 竣工年份：1994 年 /1997 年
建 设 地 点：杭州市余杭临平镇
总 建 筑 面 积 ：45000m²
总建筑高度 / 层数：99.8m/28 层
结 构 形 式：框筒结构

Client: Yuhang District Real Estate Development Company
Function: Hotel, Convention, Restaurant
Designed/Completed: 1994/ 1997
Location: Linpin, Yuhang, Hangzhou
Total Floor Area: 45,000m²
Total Height/Floor: 99.80m/ 28 Floors
Type of Structure: Framed Core-wall Structure

本工程建于杭州余杭区，是一座按四星级标准设计的集娱乐、餐饮、观光、健身、会议、住宿等功能于一体的宾馆建筑。建筑设计主要有以下特色：平面以45°斜线为母题，进行有机组合，建筑形象整体感好；三层高的自然采光中庭尺度适宜，烘托了宾馆的气氛；造型及立面设计追求简洁明快的处理手法，室内与室外，主楼与裙房比较协调。

The project is designed as 4-star grade with a 3-storey high, naturally lit atrium lobby. It is fully equipped with functions as entertainment, restaurant, health club and meeting, etc. The concept for the design seeks to show its sophisticated simplicity and coordinates the interior and the exterior.

绍兴国际大酒店

Shaoxing International Grand Hotel, Shaoxing

绍兴市国际大酒店位于绍兴市西北入口处原北海池上，酒店占地38000m²，主楼地下2层，地上23层，标准客房340间。

该建筑运用了“船”的造型构思，以体现绍兴水乡城市的传统文脉，是一家环境优美，造型美观，设施齐全的四星级涉外宾馆。

This 4-star hotel covers a total site area of 38,000 square meters, and contains 2 basement floors and 23 floors aboveground accommodating 340 guest rooms. Its ship-shaped plan acknowledges the indigenous vernacular of Shaoxing located in the region of Southern China. The fully equipped hotel is designed with inviting amenities to address the varied needs from the guests.

建 设 单 位：绍兴饭店
功 能 用 途：宾馆
设计 / 竣工年份：1993 年 /1994 年
建 设 地 点：浙江绍兴市西北入口处
总 建 筑 面 积 ：45000m^2
总建筑高度 / 层数：102m/23 层
结 构 形 式：框剪结构

Client: Shaoxing Hotel
Function: Hotel
Designed/Completed: 1990/1994
Location: North-western Gateway, Shaoxing, Zhejiang
Total Floor Area: 45,000m^2
Total Height/Floor: 102m/ 23 Floors
Type of Structure: Framed Shear Wall Structure

温州国际大酒店

Wenzhou International Hotel, Wenzhou

建 设 单 位：温州国际大酒店有限公司
功 能 用 途：宾馆
设计 / 竣工年份：1993 年 /1996 年
建 设 地 点：温州市人民路
总 建 筑 面 积 ：39500m²
总建筑高度 / 层数：100m/24 层
结 构 形 式：框剪结构

Client: Wenzhou International Hotel Co., Ltd.
Function: Hotel
Designed/Completed: 1993/1996
Location: Renmin Lu, Wenzhou
Total Floor Area: 39,500m²
Total Height/Floor: 100m/ 24 Floors
Type of Structure: Framed Shear Wall Structure

温州国际大酒店位于温州市繁华商业中心的人民路与解放街交叉口。该工程在设计中较好地处理解决了建筑基地紧迫、不规则等矛盾，设计中充分利用地形，满足了酒店建筑的使用功能。使其产生最大的经济效益。塔楼部分面对街道的交叉口为一个凹弧面，减缓了建筑对街道的压迫感，也使建筑的整个外形很有气势。整个建筑在总体设置、内部功能布局、设备硬件配置、外部造型、用材选色上都充分表现了都市酒店建筑的特征和品位。建筑平面功能、流线和分区的设计紧凑而明确、空间大小适度。

Located at the junction of the bustling Renmin Lu and Jiefang Lu, the concept design for the development maximizes efficient and effective use of this restrained and irregular-shaped site. The architectural composition is designed to follow the terrain and address the various functions of the hotel that creates its premium value. The facade of the tower building facing the junction is designed as graceful concave circular structure in order that it would not overpower the streetscape yet could have a distinctive presence as it is rising from some of the neighboring buildings. Metropolitan hotel aspects are showed in the overall site arrangement, interior function layout, service facilities, exterior finishes and material selection. All circulations and functions are well defined.

温州华侨饭店

Wenzhou Overseas Chinese Hotel, Wenzhou

建 设 单 位：杭州六合房地产开发公司
功 能 用 途：宾馆
设计 / 竣工年份：1998 年 /2003 年
建 设 地 点：温州市信河街 17 号
总 建 筑 面 积 ：47000m²
总建筑高度 / 层数：100m/24 层
结 构 形 式：框架结构

Client: Hangzhou Liuhe Real Estate Development Company
Function: Hotel
Designed/Completed: 1998/2003
Location: No.17 Xinhe Street, Wenzhou
Total Floor Area: 47,000m²
Total Height/Floor: 100m/ 24 Floors
Type of Structure: Framed Structure

温州华侨饭店地处温州市人民路与信河街的起点处。新建的温州华侨饭店，占地约 8840m²，总建筑面积 47500m²，规模为 300 间客房，内部设施按“五星”级宾馆配置。

设计力求充分体现星级饭店高雅、舒适的品质，塑造与时代气息相通的氛围，使其融入环境，成为信河街口引人注目的建筑形象。建筑顺应地形，以“板式”体块的主次组合，弧形顶部的处理，立面的垂直水平分割，并以不同的材质、色彩，打破“板式”的单一感，使建筑在体量、视觉上都达到赏心悦目。

Located at the crossroad of Renmin Lu and Jiefang Lu in the downtown of Wenzhou, the renovated 5-star hotel occupies a total site area of 8,840 square meters accommodating 300 guest rooms. The concept design strives to reflect its taste refinement, physical comfort, visual esthetics and good understanding of the immediate context that achieves a greater identity in the neighborhood. Following the terrain of the site, the goal of the design is to address the major street through the integration of bars, the curving rooftop and the vertical/linear separation on the facades, together with the use of the various materials and the expressive colors to alleviate a “wall effect” and make the building pleasant and comfortable visually and physically.

桐乡市邮电综合楼

Tongxiang Post and Telecommunication Complex, Zhejiang

桐乡市邮电综合楼地处桐乡市中心，庆丰南路与校场路交叉口西南侧，由邮政生产营业用房(5894m^2)、电信生产用房(5800m^2)和新世纪大酒店(28524m^2)三部分组成。大楼平面设计合理，交通流线清晰、便捷、安全。

建筑设计虚实对比明显、高低错落有致，处理手法简洁凝练，造型挺拔明快。顶部混凝土构架的运用不仅满足了电信通讯备用天线的设置要求，同时还与裙房构架相呼应，形成极富”动感“的光影效果，丰富了整个建筑的轮廓线，又不失现代建筑的气质与特点。该项目曾获浙江省优秀设计二等奖。

Located in the downtown of Tongxiang, the development contains three main functions: the 5,894-square-meter post offices, 5,800-square-meter telecommunication offices and 28,524-square-meter hotel. The architectural plan layout is orderly designed with a carefully manipulated circulation system. The building is design in staggered height and simple straightforward form. The concrete member covered the top of the main building addresses the demand of installing the standby antenna and echoes to the structure of the podium base. The development was awarded the honor of Second Prize of Excellent Design Zhejiang.

建 设 单 位：桐乡市邮电综合楼
功 能 用 途：宾馆、邮政营业、电信生产
设计／竣工年份：1994 年 /1997 年
建 设 地 点：浙江桐乡市庆丰南路
总 建 筑 面 积 ：40218m^2
总建筑高度／层数：89m/21 层
结 构 形 式：框剪结构

Client: Tongxiang Post and Telecommunication Complex
Function: Hotel, Post Business, Telecommunication
Designed/Completed: 1994/1997
Construction Site: Qingfeng Lu（South）, Tongxiang, Zhejiang
Total Floor Area: 40,218m^2
Total Height/Floor: 89m/ 21 Floors
Type of Structure: Framed Shear Wall Structure

杭州雷迪森大酒店

Radisson Plaza Hotel , Hangzhou

建 设 单 位：杭州国大集团有限公司
功 能 用 途：酒店
设计 / 竣工年份：1992 年 /1999 年
建 设 地 点：杭州市体育场路 333 号
总 建 筑 面 积 ：48000m^2
总建筑高度 / 层数：99.9m/25 层
结 构 形 式：框架核心筒结构

Client: Hangzhou Guoda Group Co., Ltd.
Function: Hotel
Designed/Completed: 1992/ 1999
Location: No. 333 Tiyuchang Lu, Hangzhou
Total Floor Area: 48,000m^2
Total Height/Floor: 99.90m/ 25 Floors
Type of Structure: RC Framed Core-wall Structure

杭州国大雷迪森广场酒店地处杭州市金融商贸中心，与美丽的西子湖咫尺相望。建筑集经典和时尚于一身，融传统和别致于一体。酒店拥有285间客房，其中包括45个豪华套间，同时还拥有多个风格各异、享誉杭州的餐厅和酒吧。酒店完美的会议设施，可以接待不同规格的会议和宴会。本项目的建成完善和发展了原杭州国际大厦的酒店功能，为武林商圈的繁荣作出了积极的贡献。

Located in the commercial and trade center of Hangzhou and adjacent to the scenic West Lake, the proposed development accommodates 285 guest rooms including 45 deluxe suites and restaurants and bars of varied flavors which are well-known in Hangzhou. The fully equipped convention room could host meeting and banquet of various sizes. The completion and construction of the development improves and expands the former functions of Hangzhou International Tower and should be a significant contribution to Wulin Commercial District.

富阳绿槟榔娱乐城(二期)

Green Areca Entertainment Center Fuyang (Phase II), Zhejiang

建设单位：富阳绿槟榔娱乐城
功能用途：酒店、办公
设计/竣工年份：2004年/～
建设地点：富阳迎宾北路
总建筑面积：27300m^2
总建筑高度/层数：72m/19层
结构形式：框架、核心筒结构

Client: Green Areca Entertainment Center Fuyang
Function: Hotel office
Designed/Completed: 2004/～
Location: Yingbing Lu (North), Fuyang
Total Floor Area: 27,300m^2
Total Height/Floor: 72m/ 19 Floors
Type of Structure: Framed Core-wall Structure

该工程位于富阳迎宾北路与公园路交叉口东南角,交通便利,是一个理想的酒店和办公娱乐的用地。该工程为二期。整个设计注重空间环境和视觉环境的设计，力求功能布局合理，空间变化灵活，酒店与办公互不干扰，追求生动活泼、含意深刻的建筑造型。

Located at the junction of two main boulevards of the city, the proposed project enjoys an accessible and fair-going traffic. It consists of hotel, office and entertainment functions. Incorporating spatial diversity and the visual experience, the design for the development allows for an organic function composition and high flexibility of all facilities. The hotel is definitely independent from the office building.

江苏溧阳天目湖宾馆

Liyang Tianmuhu Hotel, Jiangsu

天目湖宾馆位于江苏溧阳天目湖旅游度假区。建筑依山傍水，怀揽清秀，拥有168间豪华客房，并由国家旅游局授予五星级旅游宾馆。宾馆主体位于景区中心位置，为典型的山地建筑。建筑体形采取了东西两端逐层跌落的形象，很好地弱化了高层建筑体量上对风景区环境的冲突，丰富的天际线与周边连绵的山势亦形成良好的协调。

酒店大堂设计在基地山势的顶部，通过一个通透的玻璃过厅连接地面与大堂，宾客下车后可由自动扶梯直接进入大堂，这样的处理对山体土方破坏最少，且形成了一个别具一格的酒店入口形象。

Situated in the heart of Tianmuhu Tourist Resort, this 5-star 168-room hotel includes 8 floors constructed on a hilly site and borders the main lake with a magnificent view across water towards the hill. The complex steps in at the east and west wings that would not overpower the immediate context and coordinate into the nature environment.

The atrium lobby of the hotel is located at the top of terrain and connected to the ground floor through a glass-covered corridor. Visitors can directly enters into the lobby by escalators when they drop off, thus avoiding major earthwork and achieving a distinctive identity to the hotel entrance.

建设单位：天目湖宾馆
功能用途：五星级旅游度假酒店
设计/竣工年份：2002年/2005年
建设地点：江苏溧阳天目湖景区
总建筑面积：26000m²
总建筑高度/层数：30m/8层
结构形式：框架结构

Client: Tianmuhu Hotel
Function: 5-Star Resort Hotel
Designed/Complctcd: 2002 / 2005
Location: Tianmuhu Resort, Liyang, Jiangsu
Total Floor Area: 26,000m²
Total Height/Floor: 30m/8 Floors
Type of Structure: Framed Structure

浙江大酒店

Zhejiang Grand Hotel, Zhejiang

建 设 单 位：浙江省粮食公司
功 能 用 途：宾馆、商场
设计/竣工年份：1993 年/2000 年
建 设 地 点：杭州市武林广场南侧
总 建 筑 面 积：53000m²
总建筑高度/层数：104m/28 层
结 构 形 式：框剪结构

Client: Zhejiang Province Grain Company
Function: Hotel, Retail
Designed/Completed: 1993/2000
Location: Wulin Square（South), Hangzhou
Total Floor Area: 53,000m²
Total Height/Floor: 104m/ 28 Floors
Type of Structure: Framed Shear Wall Structure

浙江大酒店位于杭州市武林广场南侧、东临延安路。酒店主体为宾馆，裙楼设商场、娱乐及餐饮，地下层设车库及设备用房。

设计依据城市规划及西湖景观的要求，建筑体形为不规则圆弧形，面向延安路立面为三片外挑的圆弧组合，西立面为一向前伸出的小圆弧，屋顶层层跌落，以观光电梯和天线为建筑的最高点。主体建筑以镜面玻璃和铝合金板饰面，高雅纯净，富有时代感。

The hotel is composed of a hotel tower and a podium base including the retail, entertainment facilities and restaurants. Basement level accommodates the car parking and M&E rooms. Respectful of the city's planning criteria and the beauty of scenic Westlake nearby, the appearance of the tower building contains three cantilevered circular elements facing Yan'an Lu and one small protruded circular structure at the west. Roof levels step in. The main tower has facades of glass glazing and aluminum panels that forms an ensemble in sophisticated and simple architectural style, conveying a sense of times.

义乌市通信发展大楼

Yiwu Communication Development Tower, Zhejiang

建设单位：义乌市电信局
功能用途：宾馆、办公、商场
设计/竣工年份：1997年/1999年
建设地点：义乌市城区北部
总建筑面积：27000m²
总建筑高度/层数：102.9m/24层
结构形式：框剪结构

Client: Yiwu Telecommunication Bureau
Function: Hotel, Office, Retail
Designed/Completed: 1997/1999
Location: North of Downtown, Yiwu
Total Floor Area: 27,000m²
Total Height/Floor: 102.90m/ 24 Floors
Type of Structure: Framed Shear Wall Structure

义乌市通信发展大楼位于义乌市城区北部的邮电中心用地内。建筑的主要功能为电信营业、新业务演示、宾馆及办公用房、机房等。为满足建筑多功能使用及分列出入口的要求，主楼平面采用了东西各有一组交通核心的点式平面。在建筑的顶部、设置了一个四面钟楼的通讯塔架，采用了三重的方碟状造型，再配合建筑的南北向立面上大体量的斜向轮廓处理和裙房顶部的游泳池的采光顶的斜向轮廓处理，与主楼的顶部相呼应，塑造了电信建筑的明快清新的格调。该工程曾获浙江省优秀设计三等奖。

Located in the Post and Telecommunication Center near the northern downtown, the development houses the business lobby, demonstration room, hotel, office and mechanic rooms. A pair of circulation core is dedicated to the east and west side of the main tower addressing the requirement of mixed-use functions and separation of access. On the roof level a communication turret with clocks on four sides is installed and complemented to the volumetric slanted profiles on the north and south elevations of building and the clerestory above the swimming pool on the podium rooftop that echoes to the roof level of the main building. The appearance of the building exhibits its sophisticated simplicity. The project was awarded the honor of Third Prize of Excellent Design Zhejiang.

杭州西子国宾馆 8 号楼

No. 8 Building, Xizi State Hotel, Hangzhou

杭州西子宾馆新建八号楼是一组庭院建筑，用地面积为 7920m²。建筑面向西湖，与西湖南线风景区的长桥公园遥相呼应，背靠西湖著名景点雷峰夕照，整个用地环境有山有水，风景秀美。总体布置顺应用地形状，呈组团布局。建筑沿湖展开，轮廓体量错落，进退有致，沿湖有露台、游步道，还有内庭院，大部分客房有良好的面湖景观。建筑以不同的使用空间，有高有低组合造型。沿湖为1层，东南面的客房为2层。建筑以翘角四坡屋顶为主，连接建筑以坡檐平顶形式，以八角亭餐厅起翘坡顶为沿湖视觉中心，达到既统一又有变化的效果。立面用清水砖贴面为主要面材，以适当的尺度划分，配以仿石的墙裙及仿斩假石的涂料做饰边。使整个建筑体现中国传统建筑的韵味并融入湖光山色之中。

The hotel is located at the western shore of the scenic Westlake on a 7.920-square-meter site that is opposite to the Westlake and against the well-known scenic resort Leifeng Pagoda. It responds to the Long Bridge Park at a distance. Following the terrain of the site, the buildings are grouped along the lake that are scaled in height and well proportioned, and completed with terrace, the walkways and inner gardens that allows most rooms enjoy a fantastic view over the lake. Moreover the buildings are designed in "terraced" structures to meet the function requirement including one floor along the lake and two floors at the east and west. Roof level is mainly designed as warped hipped roof, yet the link building provided with pitched eaves and flat roof. Restaurant is designed as an octagonal pavilion and features warped pitched roof to enhance the visual experience towards the lake. The building design acknowledges the essence and spirit of the Chinese traditional architecture and melds it into the immediate context.

建 设 单 位：浙江西子宾馆
功 能 用 途：贵宾客房
设计/竣工年份：2002年/2004年
建 设 地 点：杭州市风景区
总 建 筑 面 积：2200m²
总建筑高度/层数：1～2层
结 构 形 式：框架结构

Client: Zhejiang Xizi Hotel
Function: VIP Guest Room
Designed/Completed: 2002 / 2004
Location: West Shore of Westlake, Hangzhou
Total Floor Area: 2,200m²
Total Height/Floor: 1～2 Floors
Type of Structure: Framed Structure

浙江金马饭店

Jinma Palace Zhejiang, Hangzhou

建 设 单 位：浙江金马饭店有限公司
功 能 用 途：涉外五星级酒店
设计 / 竣工年份：1993 年 /1995 年
建 设 地 点：萧山经济技术开发区通惠路
总 建 筑 面 积 ：50800m^2
总建筑高度 / 层数：120m/25 层
结 构 形 式：框筒结构

Client: Zhejiang Jinma Palace Co., Ltd.
Function: 5-Star Hotel
Designed/Completed: 1993/ 1995
Location: Tonghui Lu, Xiaoshan Economic Development Zone, Hangzhou
Total Floor Area: 50,800m^2
Total Height/Floor: 120m/ 25 Floors
Type of Structure: Framed Core-wall Structure

金马饭店位于萧山经济技术开发区通惠路，饭店总建筑面积为50800m^2，地下2层，地上25层，高度为120m，拥有客房425间。饭店总体设计布局合理、功能分区明确，设施齐全，装修简洁，线条明快，时代感强，是一座集旅游、购物和休闲为一体的涉外五星级旅游饭店。酒店大堂是135°的变异平面布置，三层高的共享空间，圆弧型的采光顶棚，大堂设四部自动扶梯可直达二、三层的公用空间。建筑装饰设计及用材独具匠心。饭店大堂是国家旅游局评定的“全国十大金奖大堂”之一。

The fully equipped 425-room hotel accommodates two basement floors and 25 floors aboveground. Its well-defined interior space is decorated with simple finish that conveys a sense of times. The entrance lobby of monumental proportion features a 3-storey atrium and circular clerestory that was awarded the honor of National Gold Prize of Ten Lobbies issued by the National Bureau of Tourism. From the lobby guests can ascend by four escalators to the public space on the second and third floor.

宁波盛光大酒店

Ningbo Shengguang Hotel, Ningbo

建设单位：宁波盛光包装印刷有限公司
功能用途：酒店
设计/竣工年份：2000年/2007年
建设地点：宁波市
总建筑面积：80000m^2
总建筑高度/层数：100m/26层
结构形式：框架核心筒结构

Client: Ningbo Shengguang Packing & Printing Co., Ltd.
Function: Hotel
Designed/Completed: 2000/2007
Location: High Education Zone, Ningbo
Total Floor Area: 80,000m^2
Total Height/Floor: 100m/26 Floors
Type of Structure: Framed Core-wall Structure

本工程位于宁波市高教园区内，由600间客房的五星级酒店和会议中心（为宁波高教园区配套使用）组成，主楼高100m，地下1层，地上26层，其中裙房5层，包括大堂、餐饮、娱乐、会议及屋顶泳池，六层至二十六层为客房，屋顶设一观光层。

The complex is composed of a 100-meter-high main building and 5-storey podium structure accommodating a 5-star 600-room hotel and a convention center. It is taken up by 26 floors above ground and one basement floor. From first to fifth floor are located by atrium lobby, food& beverage, entertainment, meeting room and roof swimming pool. Guest rooms are served on the sixth to 26th floor with a viewing terrace on the rooftop.

浙江海威大酒店

Haiwei Grand Hotel Zhejiang, Hangzhou

本工程地块位于杭州市滨江区行政中心以西，濒临钱塘江，居于钱江三桥与四桥之间，斜对面江岸为六和塔景区、玉皇山麓，地块规划用地55093m²，项目拟建成200m高五星级酒店及11万m²高档住宅，酒店定位侧重于商务酒店的需要，服务于滨江区大量高新企业商务活动需求，住宅设计定位于高档住宅小区，强调钱塘江的景观特色。整体造型设计力求现代、简洁、明快，住宅和酒店立面设计整体、统一。

Located at the west of Hangzhou Binjiang District Administrative Center and adjacent to Qiantang River, the proposed 55,093-square-meter site responds to the Six Harmonious Pagoda and Jade Mountain across the river. It contains a 200-meter-high hotel designed as 5-star grade and an 110,000-square-meter superior residence. The market positioning for the hotel is considered to be commercial grade serving the hi-tech enterprises in Binjiang District, while the residence to be high-grade product focusing on the landscaping importance of Qiantang River. The architectural design strives to exhibit its modern, simple and lively appearance, simultaneously embodies the unification and integration between the hotel and the residence.

建 设 单 位：浙江海威房地产开发有限公司
功 能 用 途：酒店、住宅
设计 / 竣工年份：2005 年 / 在建
建 设 地 点：浙江省杭州市滨江区
总 建 筑 面 积：210000m^2
总建筑高度 / 层数：200m
结 构 形 式：框筒结构

Client: Zhejiang Haiwei Real Estate Development Co., Ltd.
Function: Hotel, Residence
Designed/Completed: 2005 / under construction
Location: Binjiang District, Hangzhou
Total Floor Area: 210,000m^2
Total Height/Floor: 200m
Type of Structure: Framed Core-wall Structure

方远国际大酒店

Fangyuan International Grand Hotel, Taizhou

建 设 单 位：浙江方远房地产开发有限公司
功 能 用 途：宾馆、办公、娱乐、餐饮等
设计 / 竣工年份：2004 年 /2006 年
建 设 地 点：浙江省台州市
总 建 筑 面 积：93886m²
总建筑高度 / 层数：116m/28 层
结 构 形 式：框剪结构

Client: Zhejiang Fangyuan Real Estate Development Co., Ltd.
Function: Hotel, Office, Entertainment, Restaurant, etc.
Designed/Completed: 2004/2006
Location: Taizhou, Zhejiang
Total Floor Area: 93886 m²
Total Height/Floor: 116m/ 28 Floors
Type of Structure: Framed Wall Structure

项目位于浙江省台州市，为城市中心区域的重要公建和标志性建筑。建筑主要功能为五星级的宾馆、商务办公、娱乐、餐饮等。高层的点式商务楼与板式宾馆主楼连成一体，裙房及附楼沿用地的转角展开，形成了面向城市道路交叉口的开放式布局。商务主楼五层至二十层之间设有贯通的室内绿化中庭，可自然通风，既满足了部分功能用房的采光通风要求，也形成了气势宏伟的独特建筑空间。在建筑的体量处理上，主楼东侧向内收进，形成弧形收分的立面效果，表现了动感升腾的建筑气势，并通过垂直线条的强调和顶部退台的处理，使建筑体形及肌理更加丰富。屋顶采取了弧面钢结构的形式，活跃了裙房的造型又呼应了主楼的体量。

Incorporating the comprehensive program of a 5-star hotel, office, entertainment and restaurant functions, the Tower creates a dramatic new landmark on Taizhou's skyline. The high-rise business tower is integrated into the hotel main building bar with its podium base and the support building extending around the corner within the site that features an open architectural composition toward the road junction of the city. An indoor landscaping atrium passes through the 5th to 20th floor of the business main building admitting the natural ventilation and light and creating a grandiose interior space. The design to the volume suggests the east side of the main building to step in as a curved form that conveys a sense of dynamic rising, together with the accentuate to the roof lines and the design of the setback on the rooftop that makes the building rich and textured in its massing and detail. Roof level is topped with curved steel structure that enlivens the shape of the podium and echoes to the volume of the main building.

嘉兴国际中港城五星级酒店／办公楼

International Super City 5-Star Hotel/Office Development, Jiaxing

建 设 单 位：浙江盈兆房产开发有限公司
功 能 用 途：酒店、写字楼
设计／竣工年份：2003 年／~
建 设 地 点：嘉兴市中环南路
总 建 筑 面 积：120000m²
总建筑高度／层数：210m/50 层
结 构 形 式：框架核心筒结构
合作设计单位：刘荣广／伍振民建筑师事务所(香港)有限公司

Client: Zhejiang Yingzhao Real Estate Development Co., Ltd.
Function: Hotel, Office Building
Designed/Completed: 2003 / ~
Location: Zhonghuan Lu (South), Jiaxing
Total Floor Area: 120,000 m²
Total Height/Floor: 210 m/ 50 Floors
Type of Structure: Framed Core-wall Structure
In association with Dennis Lau & Ng Chun Man Architects & Engineer (H.K) Ltd.

本工程位于嘉兴市中环南路，是嘉兴国际中港城整体项目的重要组成部分。内容包括五星级酒店、办公楼以及配套服务设施。

建筑地下两层，地上50层，其中一至三层为大堂、餐饮及娱乐，四至十层为办公，十一至四十七层为客房，四十八至五十层为观光及餐厅，建筑总高度210m，气势恢宏。大楼平面为梭形，顶部以球形玻璃体镶嵌其上，宛如钻石熠熠生辉，在建筑细部的表现上，采用玻璃幕墙饰面，精致、细腻的风格给人以亲切、流畅的感觉，而整体又保持一种协调感，当夜幕降临，主体建筑挺拔俊秀，玻璃顶部更是灯火辉煌，为整个嘉兴市增添了绚丽的一笔。

The complex consists of a mega hotel/office highrise with commercial center, entertainment center and other supports functions. It includes 2 basement levels and 50 floors above ground. From the first to third floor is occupied by the lobby, restaurant and entertainment center. Offices dominate the forth to tenth floor. Hotel guest rooms are distributed on 11th to 47th floor. Sightseeing terrace and restaurant is located on 48th to 50th floor. The shuttle-shaped plan of the main building is covered by a glass sphere on the top like dazzling diamond. The facades of the building are enveloped with glass curtain wall giving a sense of sophisticated simplicity. At night the straightforward form of the building coupled with the brightened top creates a symbolic identity of the building.

常州富都香格里拉饭店

Shangri-La Traders Fudu Hotel, Changzhou

本工程位于常州新区，项目为改造工程。结合扇形的建筑平面，设计了一个螺旋上升的钢结构顶部，并增加顶部幕墙竖向线条的延伸，增强了上部形象的竖向挺拔感，打破了原建筑单一的横向肌理，横竖肌理的对比衬托，创造出一个高耸入云的标志性建筑形象。

This renovated development is proposed a spiral steel structure that reaches upward to the top level to incorporate the fan-shaped plan, while vertical line extension of curtain wall is designed on the top to reinforce the straightforward pattern of the building and compete with the linear texture of the former building that achieves a distinctive skyline symbol to the city.

建 设 单 位：侨裕集团
功 能 用 途：五星级酒店
设计/竣工年份：2003 年 /2004 年
建 设 地 点：江苏常州
总 建 筑 面 积：60000m²
总建筑高度/层数：130m/31 层
结 构 形 式：框筒结构

Client: Changzhou Qiaoyu Group
Function: 5-Star Hotel
Designed/Completed: 2003 / 2004
Location: Xinbei Development Zone, Changzhou, Jiangsu
Total Floor Area:60,000m²
Total Height/Floor: 130m/31 Floors
Type of Structure: Framed Core-wall Structure

商业建筑
Commercial Building

常州南大街步行商业中心

South Boulevard Business Center,Changzhou

常州南大街步行商业中心是常州市中心集商业零售，餐饮、娱乐、休闲、办公和公寓为一体的大型城市综合体建筑。整个项目采用步行街的构建模式，将建设场地的中心部分融入城市空间，并通过建筑退台和挑廊相结合的手法营造空中步行系统，大幅增加城市空间的容量，构建立体、复合的和休闲的购物场所。整条步行街的基本构成是一个以主轴线联系的多中心、多内核的复合式环状结构，贯穿地块中央的大“S”形步行街为主街，环绕主街的“口”字形内街为辅街，同时利用挑廊、空中连廊、屋顶花园、共享中庭等手法将底层人流引向二层，形成立体化的街道系统。地块中央设有两个中心广场，形成步行街的景观休闲中心与人流集散中心，内街串接的各个中庭则是商品流动的内核，它们环绕中心广场，赋予购物空间以不同的主题。项目建成后已成为常州市中心鲜明的地标，具有显著的社会、经济效益。

Located in the downtown of the city and incorporating the program of retail, restaurant, office, apartment buildings and recreational functions, this mix used development is designed as pedestrian street that melds the central site into the urban space and creates an aerial pedestrian circulation through the articulation of stepped in buildings and gallerias. It maximizes the city's volume and reconstructs a dynamic and multi-used recreational shopping place. Basically the site is designed in a loop completed with varied functions and connected by the main axis. The S-shaped main pedestrian street that runs throughout the center of site is articulated with the square-shaped auxiliary inner street that is around the main street. Pedestrian circulation is directed to the second floor through the galleria, aerial corridor, roof garden and the atrium lobby. Two central plazas are located in the center of the site acting as a planted recreational center and concourse. Each atrium connected by the inner street serves as the hub of delivery and encloses the central plazas to give the shopping space different theme. The design for the development creates a dramatic new landmark to the city under completion and has made contribution to the economy and society.

建设单位：常州之江置业有限公司
功能用途：大型城市商办住综合体
设计/竣工年份：2002年/2005年
建设地点：常州市南大街
总建筑面积：210000 m^2
总建筑高度/层数：96.50m/ 31层
结构形式：框架结构、框架剪力墙结构

Client: Changzhou Zhijiang Real Estate Co., Ltd.
Function: Pedestrian Street, Business, Office, Housing
Designed/Completed: 2002/2005
Location: South Boulevard, Changzhou
Total Floor Area: 210,000m^2
Total Height/Floor: 96.50m / 31 Floors
Type of Structure: Framed & Framed Shear Wall Structure

上海龙华旅游城

Longhua Tourist Town,Shanghai

龙华旅游城位于上海徐汇区南部的龙华镇，与龙华寺、龙华烈士陵园和筹建中的上海历史博物馆共同组成上海龙华游旅区。1996 年龙华旅游城被国家旅游总局列为“九五”期间上海市区五大景区之一。

旅游城总建筑面积 6 万 m^2，集食、住、行、游、购、娱等功能于一体。建筑虽为钢筋混凝土结构，取江南明清时期水乡民居风格，粉墙黛瓦，古今并蓄，以新的面貌传递古风。

The development combines Longhua Temple, Longhua Cemetery of Revolutionary Martyrs and the under constructed Shanghai Historical Museum into the Shanghai Longhua Tourist Zone. In 1996 the development is listed as one of the five resorts in Ninth Five-year Plan by National Tourist Bureau. It accommodates restaurant, hotel, tourism and retail functions. The architectural building acknowledges the indigenous vernacular in Ming and Qin Dynasty coupled with white-washed wall and black clay tile that is one of modern interpretation of traditional Chinese architecture.

麦当劳

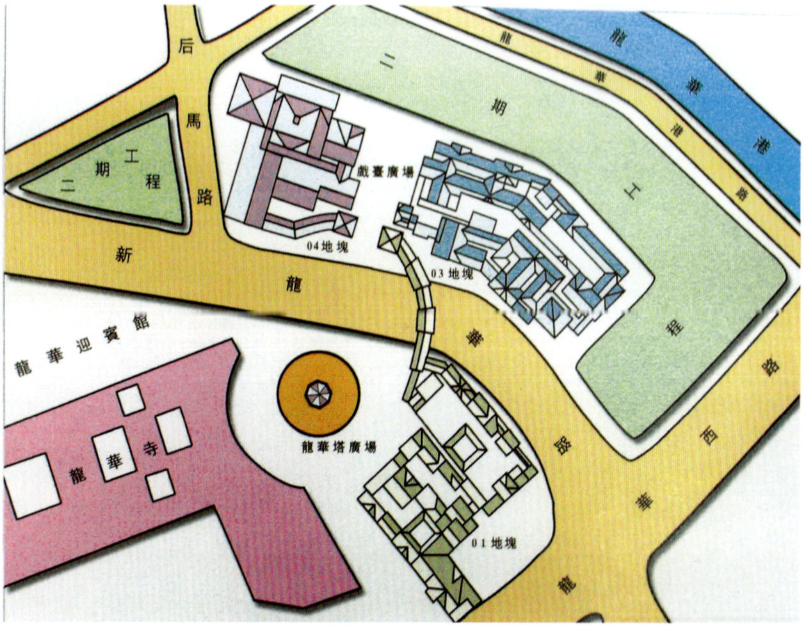
后馬路
二期工程
新龍華路
龍華港
龍華港路
二期工程
戲臺廣場
04地塊
03地塊
龍華迎賓館
龍華寺
龍華塔廣場
01地塊
龍華西路

鬧市大吉

建设单位：上海龙华旅游城开发公司
功能用途：商业
设计/竣工年份：1999年/2000年
建设地点：上海市徐汇区南部龙华镇
总建筑面积：60000m²
总建筑高度/层数：17m/3层
结构形式：钢筋混凝土结构

Client: Shanghai Longhua Tourist Town Development Corp.
Function: Commercial
Designed/Completed: 1999/2000
Location: Longhua Town, Xuhui District, Shanghai
Total Floor Area: 60,000m²
Total Height/Floor: 17m/ 3 Floors
Type of Structure: RC Structure

无锡通惠路商业步行街

Tonghui Lu Commercial Pedestrian Street, Wuxi

通惠路商业街位于无锡市城北原通惠路两侧，计划建成大型城市商贸综合体群落。整个规划设计的基本构思立足于利用地块便利的交通条件，以及处于城市副中心的优越地理位置，克服地块纵深过大，周边商业设施档次较低的不利因素，力求以完整、成熟的商业业态群与顺畅的内部交通组织，创造新型且成规模化的商业群体，形成无锡市的商业副中心区。设计以划分和构筑多主题的步行街坊群为起点，在各群体内建造立体化的购物空间，并将线形街道与环形绿化景观空间相结合，组织水街、下沉广场、多层景观平台等休闲区域，通过活跃的市民活动形成多变而精彩的购物氛围，实现城市改造的根本目标。

The program for the project is proposed to incorporate the department store, supermarket, retail shop, restaurant, hotel and housing into a comprehensive development. The concept design for the project aims to maximize the use of the efficient transportation, minimize the disadvantage of the depth of the site and improve the business functions around it. Besides the development is to be provided with completed mature commercial facilities and fair-going inner traffic that becomes a modern and subsidiary business center of the city. The design starts from the principle of separating and constructing multi-themed neighborhoods to create a multi-storey shopping space in each neighborhood and integrate the linear street into the looped landscaping space in which to incorporate the water element, sunken plaza, multi-storey terrace and other recreational facilities. It strives to create a diversified and wonderful shopping environment related to the civic activities that realizes the principal objective of urban rejuvenation.

水街休闲区内景透视图

时尚精品区内街透视

大众零售区下沉广场透视

建 设 单 位：无锡金太湖房地产开发有限公司
功 能 用 途：大型城市商办住综合体
设计 / 竣工年份：2004 年 / 在建
建 设 地 点：无锡市通惠路
总 建 筑 面 积 ：483300 m^2
总建筑高度 / 层数：175m/ 40 层
结 构 形 式：框架结构、框架剪力墙

Client: Wuxi Gold Tai Lake Real Estate Development Co., Ltd.
Function: Commercial, Office, Residence
Designed/Completed: 2004 / under construction
Location: Tonghui Lu, Wuxi
Total Floor Area: 483,300m^2
Total Height/Floor: 175m/ 40 Floors
Type of Structure: Framed & Framed Shear Wall Structure

杭州西城广场

Season Square, Hangzhou

建 设 单 位：杭州新西元置业有限公司
功 能 用 途：商场、餐饮、休闲、健身
设计 / 竣工年份：2003 年 /2004 年
建 设 地 点：杭州城西蒋村中心
总 建 筑 面 积 ：55000m²
总建筑高度 / 层数：24m/6 层
结 构 形 式：框剪结构

Client: Hangzhou Xinxiyuan Real Estate Development Co., Ltd.
Function: Retail, Restaurant, Entertainment Facilities, Health Club
Designed/Completed: 2003/2004
Location: Jiang Cun, West of Hangzhou
Total Floor Area: 55,000m²
Total Height/Floor: 24m/ 6 Floors
Type of Structure: Framed Shear Wall Structure

西城广场位于杭州市城西蒋村中心位置，广场占地面积 32300 m²，是杭州目前唯一将购物中心与大型景观广场紧密结合的集购物、美食、娱乐、休闲、健身、观光于一体的大型综合性全景商业建筑中心，也是杭州唯一自带 18000m² 大型绿地休闲广场的商业物业。

Located at the west of Hangzhou, the development occupies a site area of 32,300 square meters. It is the only multi-functional commercial center in Hangzhou that integrates the shopping mall into the great landscaping plaza and combines the retail, food & beverage, entertainment facilities, health club and tourist destination. It is also the only commercial property with a unique open landscaping plaza of 18,000 square meters.

season square

免费停车场
FREE PARKING
KFC

厦门聚泰商业城

Jutai Commercial Town, Xiamen

建设单位：厦门聚泰房产开发有限公司
功能用途：住宅、商场、写字楼
设计/竣工年份：1991年/1993年
建设地点：福建厦门展览广场
总建筑面积：32000 m^2
总建筑高度/层数：82m/22层
结构形式：框剪结构

Client: Xiamen Jutai Real Estate Co., Ltd.
Function: Residence, Retail, Office
Designed/Completed: 1991/1993
Location: Exhibition Square, Xiamen, Fujian
Total Floor Area: 32,000m^2
Total Height/Floor: 82m/ 22 Floors
Type of Structure: Framed Shear Wall Structure

聚泰商业城位于厦门商业中心地段，地下二层为设备配套用房，地下一层设大型双层机械式停车库，底层至四层为大型商场，五至八层为小间式写字楼，九层以上为住宅公寓。

因该楼以住宅公寓为主要内容，因此建筑为简洁的板式造型，但结合功能布置特点，将阳台、门窗作有机的组合，创造出板式建筑的新形象，建筑色彩大胆的选择了红、白两色相间的三段式形象，成为该地段内颇具特色的新建筑。

Located in the bustling downtown of the city, the multi-functional development contains two basement levels providing mechanic and electrical rooms on Basement Level 1 and dual-level automatic car parking on Basement Level 2. From the ground to the forth floor is designated by the large department store. Cellular offices are housed on the fifth to eighth floor. Above the ninth floor is the apartment building. As the main component of the project, the apartment building is designed in the form of simple bar but integrated with the interior functions by articulating the balconies, windows and doors that alleviates the "wall effect" towards the street. The symbolization and significance of the design is defined by the bright colors applied to the facades.

杭州湖滨路商贸旅游特色街区 15 号地块工程

Hangzhou Shore District Development Lot 15, Hangzhou

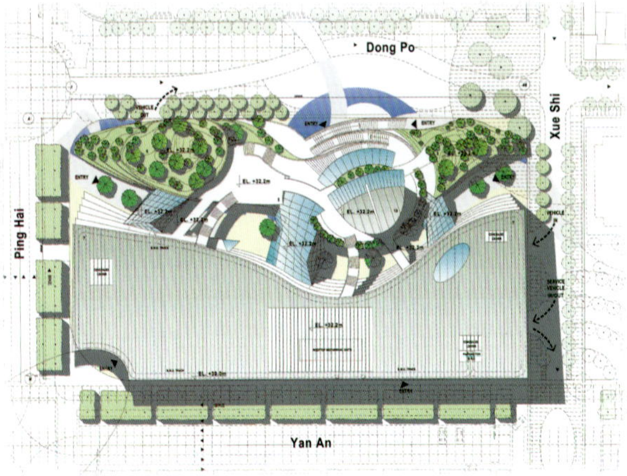

建 设 单 位：杭州湖滨地区特色街居建设有限公司
功 能 用 途：商业
设计 / 竣工年份：2003 年 /2006 年
建 设 地 点：杭州市湖滨地区
总 建 筑 面 积：65000m²
总建筑高度 / 层数：26m/9 层（其中地下 2 层）
结 构 形 式：框架结构，局部钢结构
合作设计单位：美国捷得国际建筑事务所(JERDE)

Client: Hangzhou Waterfront Vernacular District Development Co., Ltd.
Function: Commercial
Designed/Completed: 2003/2006
Location: Shore District, Hangzhou
Total Floor Area: 65,000m²
Total Height/Floor: 26m/ 9 Floors (including 2 floors below ground)
Type of Structure: Framed & Steel Structure
In association with The Jerde Partnership International. Inc.

湖滨地区 15 号地块工程项目是湖滨地区商贸旅游特色街区建设整治工程的一个组成部分，地块位于延安路、平海路、东坡路和学士路所围合的街区内。项目旨在使之成为湖滨地区具有鲜明西湖地域特征的现代旅游商贸场所。设计创造了一个“岛”和“峡谷”的构思来体现建筑丰富多样的空间，为人们提供了多层次的奇妙的场所体验。地块建筑的主要功能内容为地下停车场、商场、商铺、休闲餐饮和小影视城。

As one of the main conceptual components of the master plan and the associated Lots, Lot 15, the Canyon is located on the intersection of Yan'an Road on the east, and Pinghai Road on the south, and is flanked on west by Dongpo Road and Xueshi Road on the north. The architectural expression of the project along the two lanes landscaped, one-way Dongpo Road is organic in image and "island like" in detail and massing. The canyon creates an opportunity to express a dynamic opposition on the canal, an organic island component on the Lakeside and an urban architectural component on the City side. The project has many fascination layers of experience, adventure and embodies a profound sense of place. The canyon is a memorable addition to the Westlake atecture district.
The project has one full retail level and one parking level below grade, and also includes kiosk, restaurant, mini cinema and other entertainment facilities.

杭州湖滨路商贸旅游特色街区19号、22号、25号地块工程

Hangzhou Shore District Development Lot 19 22 25, Hangzhou

湖滨地区19号、22号、25号三个地块工程项目是湖滨地区商贸旅游特色街区建设整治工程的一个组成部分，地块位于吴山路、解放街、延安路、平海路所围合的街区内，被邮电路、仁和路分为三个地块。这一地块是连接延安路一侧的现代新型商店和办公建筑，以及吴山路一侧的传统韵味的里弄和其他建筑的重要中介空间，这种中介空间、街道或称作共享空间是整个设计的命脉和灵魂所在，它的灵感来源于传统园林的"曲径通幽"，同时采用一种现代和高科技的方式表达，其特征为既具有现代风格，又保留、体现历史遗风的步行内街多层建筑。建筑的主要功能为地下停车场、商场、商铺、休闲餐饮、写字间工作室等。

Lot 19 22 25 are located on Yan'an Road to the west and Wushan Road to the east with Jiefang Road to the south and Pinghai Road to the north and divided into three lots by Youdian and Renhe Road. The project is in fact the critical interface between the modern new architecture of the retail and studio buildings on Yan'an Road, and the historical reconstruction of the shop houses and period buildings along Wushan Road. The configuration of this in between space, "the street", is where it all happens, it is the life and soul of the project. It is inspired by the 'crooked court yard' path and yet, expressed in a modern and technically advanced way. This configuration will develop into an extremely interesting and exciting space, expressing the dynamics of the program and architectural typologies on either side. The overall program for this street portion of the district generally includes two levels of retail, one level of restaurants and two levels of studio offices spaces.

建 设 单 位：杭州湖滨地区特色街居建设有限公司
功 能 用 途：商业
设计 / 竣工年份：2002 年 / 2006 年
建 设 地 点：杭州市湖滨地区
总 建 筑 面 积： 146400m²
总建筑高度 / 层数：25m / 7 层（其中地下 2 层）
结 构 形 式：框架结构，局部钢结构
合作设计单位：美国捷得国际建筑事务所(JERDE)

Client: Hangzhou Waterfront District Development Co., Ltd.
Function: Commercial
Designed/Completed: 2002 / 2006
Location: Shore District, Hangzhou
Total Floor Area: 146,400m²
Total Height/Floor: 25m/ 7 Floors (including 2 floors below ground)
Type of Structure: Framed & Steel Structure
In association with The Jerde Partnership International. Inc.

杭州吴山花鸟市场

Wushan Flower and Bird Market, Hangzhou

建 设 单 位：杭州吴山商城开发有限公司
功 能 用 途：商业
设计 / 竣工年份：2001 年 /2002 年
建 设 地 点：杭州河坊街西端
总 建 筑 面 积 ：28000m²
总建筑高度 / 层数：21m/6 层
结 构 形 式：框架结构

Client: Hangzhou Wushan Commercial Center Development Co., Ltd.
Function: Commercial
Designed/Completed: 2001/2002
Location: Hefang Street (West), Hangzhou
Total Floor Area: 28,000m²
Total Height/Floor: 21m/ 6 Floors
Type of Structure: Framed Structure

杭州吴山花鸟城是集商场和餐饮为一体的综合性商业建筑。

由于吴山花鸟城地处吴山广场辐射区域内，且位于保留的河坊街（步行街）的最西端，地理环境独特，对建筑设计也提出了独特的要求。整个建筑在造型上重心东移，使得建筑物自西而东逐层后退，在体量上能和西侧的吴山广场的开阔空间相呼应，也和南侧的吴山山形相一致。设计大量采用传统建筑特征的建筑构件和符号以及现代建筑材料和一系列悬挑式平台，使该建筑成为与环境相协调，具有民族韵味的现代化建筑。

Located at the west terminal of the historical Hefang Pedestrian Street within Wushan Square, the mixed-used development, including retail and restaurant functions, enjoys a splendid location and should be designed for maximum efficiency to comply with the City's planning criteria. The buildings are stepped in from the west to the east that echoes to the open Wushan Square at the west side and corresponds to the hilly terrain of Wushan Mountain at the south. With its numerous symbolic traditional architectural elements to be incorporated with the new materials and the cantilever terraces, the modern building architecture melds into the immediate context and acknowledges the indigenous vernacular.

杭州吴山商城

Wushan Business Town, Hangzhou

建 设 单 位：杭州吴山商城开发有限公司
功 能 用 途：商贸、住宅、办公
设计 / 竣工年份：2001 年 /2003 年
建 设 地 点：杭州市吴山广场北侧
总 建 筑 面 积 ：95000m²
总建筑高度 / 层数：24m/7 层
结 构 形 式：框架结构

Client: Hangzhou Wushan Commercial Center Development Co., Ltd.
Function: Business, Residence, Office
Designed/Completed: 2001/2003
Location: Wushan Square （North）, Hangzhou
Total Floor Area: 95,000m²
Total Height/Floor: 24m/ 7 Floors
Type of Structure: RC Framed Structure

吴山商城地处杭州著名的吴山风景区，是一个集商业、办公、住宅于一体的综合性建筑群体，它由二座商务楼和一幢高级住宅楼组成。

商务楼的主要特征在于它拥有较宽敞的双层共享中庭空间，通过优化的室内设计，使之堪称“可自然呼吸的绿色中庭”。

建筑采取了退台式的造型设计，使得建筑体形与吴山景区相协调。

The development is composed of 2 business office buildings and one high quality residential building. The business office buildings is dominated by two-storey atrium lobby with well-proportioned interior to be designed as a “landscaping atrium allowing for naturally breathing”. Due to its splendid location, the complex steps in to correspond to the scenic resort and to provide spacious terraces to the offices.

常州大学城中央商业街

Central Commercial Street Changzhou University Town,Changzhou

F区透视

该项目位于常州武进区南部,路北为由六所大学组成的常州大学城，南为大学新村,是武进区重要的景观道路之一。项目是未来城市发展的核心商业区，集商贸、商务、购物、餐饮、休闲、娱乐、酒店、公寓为一个统一的综合体。

Predominantly located next to a landscaped boulevard in Wujin District, the development is bounded by Changzhou University Town to the north which includes six universities and the newly built University New Community in the south. It would be the entertainment and commercial hub of the city occupied by trade, business, retail, restaurant and entertainment functions besides the hotel and apartment building.

建 设 单 位：常州市武进区建设局
功 能 用 途：商业
设计 / 竣工年份：2004 年 /2005 年
建 设 地 点：常州市鸣新中路
总 建 筑 面 积 ：180000m²
总建筑高度 / 层数：24m/6 层
结 构 形 式：框架结构

Client: Construction Bureau of Wujin District, Changzhou
Function: Commercial
Designed/Completed: 2004/2005
Location: Mingxin Zhong Lu, Changzhou
Total Floor Area: 180,000m²
Total Height/Floor: 24m/ 6 Floors
Type of Structure: Framed Structure

泰州坡子街商业中心

Puo Zi Street Business Center, Taizhou

泰州市坡子街商业中心位于泰州市旧城区核心地段，是泰州市的传统商业街区，在苏中、苏北地区曾经极具影响力和辐射力，近年来因为种种原因，该区域处于日渐衰退的过程中。本项目顺应地块南面月城广场的改造和东面滨河绿地公园的兴建，以及周边城市道路的拓宽整治，力求以完备的功能配置、整合的商业业态布局、优化的动静态交通条件和休闲化的购物氛围来重振城市活力。整个项目的中心地块取“回”字型结构布置，坡子街作为核心区域被改造为带大型下沉广场的步行主街，周边建筑的分层退台赋之以丰富的空间感受；环绕下沉广场外圈的步行外街串连着四个小型主题广场，满足了业态布局可分可合的要求。建筑外形简洁现代，暖色调的色彩运用体现商业建筑的个性特点。高层公寓和住宅沿建筑群北侧布置，保证了商业街区内良好的尺度控制和充足的日照条件。

Located in the hub of the old city of Taizhou, the development is within the traditional business district. To correspond to the renovation of the Yuecheng Plaza at south, the redevelopment of the riverbank green park at east and the expansion of the immediate urban roads, this mix-used development is designed with fully equipped facilities, integrated business functions, well-defined circulation system and welcoming shopping amenities to liven up the district which used to have great influence and importance in the middle and the north region of Jiangsu Province but gradually became declined. The overall site planning arrangement is designed in 回(one of Chinese character)shape. Puozi Street, as the center of the site, is designed as the pedestrian street with the sunken plaza of monumental proportion and surrounded by the buildings which step in toward it to achieve spatial diversity and formal richness. The outer pedestrian street around the plaza connects four small themed plazas allowing the high flexibility of all facilities. The simple form of the building coupled with the warm colors refers to the essence of the business architecture. High-rise apartment and residential buildings are located along the north of the ensemble that achieves the optimal scale and day light to the site.

建 设 单 位：泰州宝江置业有限公司
功 能 用 途：商业步行街、酒店式公寓与住宅综合体
设计 / 竣工年份：2003 年 / ~
建 设 地 点：泰州市坡子街
总 建 筑 面 积：$223700m^2$
总建筑高度 / 层数：80m/26 层
结 构 形 式：框架结构、框架剪力墙

Client: Taizhou Baojiang Real Estate Co., Ltd.
Function: Pedestrian Street, Service Apartment, Residence
Designed/Completed: 2003 / ~
Location: Puozi Street, Taizhou
Total Floor Area: $223,700m^2$
Total Height/Floor: 80m/ 26 Floors
Type of Structure: Framed & Framed Shear Wall Structure

上海新伟商务大厦

Xinwei Business Tower, Shanghai

上海新伟大厦位于普陀区新兴的商业中心内，紧邻外环高架和沪宁高速出入口。该项目由5幢总计10万m^2的高层办公建筑和总计28万m^2的7层商业裙房组成。整个建筑群落及裙房部分顺应地形呈“L”形布局，屋顶上设空中庭院，布置了游乐场、游泳池及屋顶花园，高层建筑则沿周边布置。同时，设计上以“S”形的步行购物街串连一、二期建筑，创造人性化的购物空间，顶部以半封闭的玻璃顶遮盖，保证建筑通风效果。整个建筑群落的布局充分考虑了与周边建筑的关系，以及自身天际线，在城市范围内形成标志性的优美景观。建筑外观设计简约明快，通透亮丽，展现出现代商业大都会的建筑风范。

Located in the new business center of Putuo District and near the outer ring freeway and the exit to Shanghai-Ningbo Express, the complex consists of five high-rise office buildings of 100,000 square meters and 7-storey business podium structure of 280,000 square meters. The podium base is arranged on the L-shaped site and provided with rooftop courtyard occupied by the playground, swimming pool and roof garden, while the high-rise tower is situated at perimeter of the site. The S-shaped pedestrian shopping street which runs through Phase One and Phase Two is covered with the semi enclosed glass canopy that achieves effective and efficient ventilation. The appearance of the building is simple in its form and transparent in its volume that exhibits the modern architectural style of metropolitan building.

建 设 单 位：上海新伟置业有限公司
功 能 用 途：大型城市商办综合体
设计 / 竣工年份：2004 年 / 在建
建 设 地 点：上海市普陀区真北路 / 梅川路
总 建 筑 面 积 ：500000m²
总建筑高度 / 层数：99.75m/ 25 层
结 构 形 式：框架结构、框架剪力墙结构

Client: Shanghai Xinwei Real Estate Co., Ltd.
Function: Commercial, Office
Designed/Completed: 2004 / under construction
Location: Zhenbei Lu, Meichuan Lu, Putuo District, Shanghai
Total Floor Area: 500,000m²
Total Height/Floor: 99.75m/ 25 Floors
Type of Structure: Framed & Framed Shear Wall Structure

文教建筑
Cultural and Education Building

复旦大学上海视觉艺术学院

Shanghai Institute of Visual Art, Fudan University, Shanghai

复旦大学上海视觉艺术学院总用地面积为50hm^2，总建筑面积为133000m^2，位于上海松江大学城。设计规模为学生3000人，教师员工1000人。作为全国第一所视觉艺术大学，其建筑造型风格独具个性。各单体建筑形态各异，和谐而多变，像一组精心设计的雕塑散落在优美的自然景观中，给人以强烈的艺术感染和各不相同的视觉联想。主要建筑物有：展示图文中心、行政中心空间学部、时尚学部、公共教学楼、美术学部、设计学部、影视学部、媒体艺术学部等。

The development covers an overall site area of 50 hectares accommodating 3000 pupils and 1000 staffs. As the first institute for visual art in China, the building architecture is unique in its form. Each block is designed in varied styles but in harmonious manner recalling the representative sculptures dispersed in the natural environment. The expressiveness creates a strong artistic appeal and varied visual association. The main functions within the site include the exhibition and library center, administrative center, school of spatial and industrial design, school of fashion design, public teaching building, school of fine arts, school of design, school of film & TV production and school of media and art.

建 设 单 位：复旦大学上海视觉艺术学院
功 能 用 途：教学
设计 / 竣工年份：2003 年 /2005 年
建 设 地 点：上海松江大学城
总 建 筑 面 积 ：133000m^2
总建筑高度 / 层数：24m/4～5 层
结 构 形 式：框架结构

Client: Shanghai Institute of Visual Art, Fudan University
Function: Education
Designed/Completed: 2003/ 2005
Location: Songjiang University Town, Shanghai
Total Floor Area: 133,000m^2
Total Height/Floor: 24m/ 4～5 Floors
Type of Structure: Framed Structure

浙江工业大学屏峰校区西组团

West Parcel of Pingfeng Campus,Zhejiang Industrial University, Hangzhou

本地块位于校区西侧。遵循原校区总体规划中对各学院组团的总体布局，东侧为经贸管理学院，西侧由南向北分别布置了人文学院、法学院、政管学院和国际学院。五大学院相互围合形成一个大型的中央绿化，其下部为大型地下停车场。在总图布局中将经贸管理学院往西偏移，使之与入口大道之间形成一个60m宽的绿化带，尽显大气风格。西侧人文学院、法学院、政管学院和国际学院的平面布局均采用庭院式风格，与整个校区东侧已建教学楼布局手法相一致。

The site is situated at the west of the campus. Respectful of the overall master plan to each academic parcel within the site, the School of Economic Trade Management is located at the east and the west from the south to the north is dominated by the School of Humanities, School of Law, School of Political Administration and International School encompassing a spacious planted garden with basement car parking under it. The overall design approach proposes the School of Economic Trade Management moved a little to the west that creates a 60-meter-wide green land between it and the entrance road. The architecture for the four schools located at the west is designed in the form of quadrangle that corresponds to the built teaching buildings at the east of the campus.

建设单位：浙江工业大学
功能用途：教学楼
设计/竣工年份：2004年/2005年
建设地点：杭州小和山
总建筑面积：6644/m^2
总建筑高度/层数：4～5层
结构形式：框架结构

Client: Zhejiang Industrial University
Function: Teaching Building
Designed/Completed: 2004/2005
Location: Xiaohe Mountain, Hangzhou
Total Floor Area: 66,447m^2
Total Height/Floor: 4～5 Floors
Type of Structure: Framed Structure

上海对外贸易学院松江校区

Songjiang Campus, Shanghai University of Foreign Trade, Shanghai

上海对外贸易学院松江新校区位于上海松江大学城，建筑用地面积为39.35hm²。学院总体布局、景点的分布、建筑的造型等方面均以营造新世纪现代高等院校的环境特色为出发点，力求形成一个校园建筑、绿化景点与自然环境融为一体的校园景观。

校前区的主要建筑为语言信息中心和教学楼。在语言信息中心大楼南立面的正中设计了一个大型拱门墙。拱门中为大型台阶。教学楼的西边，设有纵贯南、北的大型山墙，墙上也设有三个圆拱门，每一圆拱下都配以大台阶。这一圆拱长墙与语言信息中心的圆拱南墙相互衬映，形成宏伟的气势，在景观上具有“镇校”的突出效果。

Located in Songjiang University Town Shanghai, the proposed development occupies an overall site area of 393,500 square meters. The design of the general plan layout, the landscaping settings and the shapes of buildings are initiated with the concept of creating a unique environment to the state-of-the-art high education campus in the new century that melds the architecture, the planted land and the natural environment into a single, comprehensive expression.The pre-functions include the Language Information Center and Teaching Buildings. In the core of the south elevation of the Language Information Center is featured a grand arch with grand steps in the middle. At the western teaching building is designed with a grand gable wall through the south to the north and completed with three arches on the wall. It corresponds to the southern arch of the Language Information Center that appears as a representative sculpture and creates an imposing ensemble to the campus.

建 设 单 位：上海对外贸易学院
功 能 用 途：教学
设计 / 竣工年份：2000 年 /2003 年
建 设 地 点：上海松江大学城
总 建 筑 面 积 ：105000m²
总建筑高度 / 层数：23.80m/5 层
结 构 形 式：框架结构

Client: Shanghai University of Foreign Trade
Function: Education
Designed/Completed: 2000/2003
Location: Songjiang University Town, Shanghai
Total Floor Area: 105,000m²
Total Height/Floor: 23.80m/ 5 Floors
Type of Structure: Framed Structure

浙江水利水电学校

Zhejiang Water Conservancy And Hydroelectric Power School, Hangzhou

项目位于杭州市萧山区钱江科教城内，地块北侧为科教城内规划主干道，西侧为规划二路，南侧为先锋河，规划用地面积为 133847m^2。用地呈长方形，东西宽约230m，南北长约550m，项目建设内容包括教学楼群，图书信息中心、继续教育中心、实验楼、实训楼、学生公寓、行政楼、食堂、文体活动中心、单身宿舍等，总建筑面积102303 m^2，建成后将成为容纳 3500 名在校生的现代化院校。

The development occupies an overall site area of 133,847 square meters. The main functions within the rectangular-shaped site, 230 meter wide from the east to the west and 550 meter long from the south to north, include the teaching building complex, library/information center, continuous learning center, laboratory, practice/training building, apartment building for students, administrative building, canteen, recreational / sports center and dormitory for the bachelor, etc. It would accommodate 3500 students upon completion.

建 设 单 位：浙江水利水电学校
功 能 用 途：教学、培训
设计 / 竣工年份：2005 年 /2006 年
建 设 地 点：杭州市萧山区钱江科教城
总 建 筑 面 积：102303m²
总建筑高度 / 层数：2～8 层
结 构 形 式：框架结构

Client: Zhejiang Water Conservancy And Hydroelectric Power School
Function: Education, Training
Designed/Completed: 2005/2006
Location: Qiantang River Scientific Education Town, Xiaoshan, Hangzhou
Total Floor Area: 102,303m²
Total Height/Floor: 2～8 Floors
Type of Structure: Framed Structure

南京体育学院仙林分校

Nanjing Sports Institute Xianlin Branch, Nanjing

本工程位于南京市仙林大学城内，占地66.7hm^2，属丘林地貌，是南京体育学院的分校，属竞技体育校区。规划设置28个运动项目，运动员1500人，教职员工500人。工程分二期，一期工程内容有乒乓馆、羽毛球馆、篮排球馆、田径馆、举重馆、武艺馆、击剑馆、体能训练馆、自行车赛场、教学实验楼、科研综合楼、办公楼、图书馆、保健中心、运动员公寓、教职工宿舍、运动员、学生及职工餐厅、冠军楼、后勤用房及室外田径场（2片），沙滩排球场（4片），二期工程有网球馆、游泳馆、水球馆、体操馆、手球馆、柔道馆、运动员俱乐部、网球场等。其中自行车赛场为第十届全国运动会的正式比赛场。

The development covers a total site area of 66.7 hectares and accommodates 28 sport events, 1500 athletes and 500 coaches and staffs. It is divided into two phases. Phase One includes table tennis hall, badminton hall, basketball hall, volleyball hall, track and field hall, weight lifting hall, wushu hall, fencing hall, physical and skill training hall, cycling track, teaching and laboratory building, scientific research building, administrative office, library, healthcare center, apartment building for athletes, dormitory for coach and staff, cafeteria, building for champions, back-of-house, two outdoor tracks and fields and four beach volleyball courts. Phase Two contains tennis hall, natatorium, water polo hall, gymnastic hall, handball field, judo hall, club for athletes and tennis court. The cycling track would be designed as the formal venue for the Tenth National Game.

建 设 单 位：南京体育学院仙林分校建设办
功 能 用 途：训练、教学、比赛
设计 / 竣工年份：2002 年 /2005 年
建 设 地 点：南京市仙林大学城内
总 建 筑 面 积 ：192000m²
总建筑高度 / 层数：23m/1～6 层
结 构 形 式：框架、砖混、框架＋钢屋架，框架＋膜结构

Client: Task Force of Nanjing Sports Institute Xianlin Branch
Function: Training, Education, Competition
Designed/Completed: 2002/2005
Location: Xianlin University Town, Nanjing
Total Floor Area: 192,000m²
Total Height/Floor: 23m/1～6 Floors
Type of Structure: Framed Structure, Concrete & Brick, Steel Roof Truss, Membrane

浙江工商大学下沙校区

Xiasha Campus of Zhejiang Gongshang University, Hangzhou

作为一个新型校园的规划设计，其总体布局倡导具有时代气息的校园文化，充分体现了自身特色。设计通过两条轴线，一个步行系统与一条机动车环路，三条绿化景观带的设置布局，达到室内外空间环境的渗透及景观视线的穿插。

教学群体建筑集中布置，成"船形"几何体构图，错落有致，从低到高，由东西两翼逐渐汇集中心，烘托、突出了主体建筑对全局的控制地位。

The master plan concept presents a state-of-the-art campus that responds to the essence of the building and conveys a sense of times. In the overall site arrangement, the design proposes two axis, a pedestrian circulation, a vehicular loop road and three landscaping strips to create the interpenetration of interior and exterior space and interweave between the landscaping sightlines. Teaching buildings are grouped in the form of ship and staggered in height focusing on the predominant location of the main building.

墨湖

圖書館

建 设 单 位：杭州商学院
功 能 用 途：教学
设计 / 竣工年份：2003 年 /2004 年
建 设 地 点：杭州市下沙
总 建 筑 面 积 ：300000m²
总建筑高度 / 层数：52m/12 层
结 构 形 式：框架结构

Client: Zhejiang Gongshang University
Function: Education
Designed/Completed: 2003/2004
Location: Xiasha, Hangzhou
Total Floor Area: 300,000m²
Total Height/Floor: 52m/ 12 Floors
Type of Structure: Framed Structure

上海师范大学奉贤校区

Shanghai Teacher's University Fengxian Campus

上海师范大学奉贤校区位于奉贤海湾旅游开发区，规划用地面积83.3hm^2。建有图文信息中心、教学楼、行政楼、交流中心、体育中心等单体建筑。整个校园以中央圆形广场为视觉中心，并以系列的轴线空间构图把各单体部分有机的联系起来。外观造型采用四坡屋顶形式来统一整个校园的建筑主导风格，舒缓的坡顶，创造了一种田园式的宁静悠远的教育环境氛围，同时与绿水、草地、小丘相得益彰，构成了一幅优美的画卷。

The development occupies a total site area of 83.33 hectares and houses the audio/visual information center, teaching building, administrative building, communication center and sports center. A round square is predominantly situated in the center of the campus. All functions are connected in an organic configuration through a series of axis. Hipped roofs are designed to unify the overall architectural style in the campus and complemented with the surrounding water features, grass and hills recalling a pastoral serene and pleasant environ.

建设单位：上海师范大学
功能用途：教学
设计/竣工年份：2001年/2003年
建设地点：上海奉贤海湾旅游开发区
总建筑面积：82000m²
总建筑高度/层数：19m～36m/3～9层
结构形式：框架结构

Client: Shanghai Teacher's University
Function: Education
Designed/Completed: 2001/2003
Location: Fengxian Gulf Tourist Development Zone, Shanghai
Total Floor Area: 82,000m²
Total Height/Floor: 19m～36m/ 3 ～9Floors
Type of Structure: Framed Structure

上海大学新校区

Shanghai University New Campus, Shanghai

上海大学新校区位于上海市西北宝山大场镇范围内，基地总面积101hm²，分三期建设完成。新校区包括美术学院、文学院、外语学院、国际商学院、社会科学学院、理学院、计算机学院、法学院、知识产权学院及影视制作中心等。

规划布局及建筑设计主要考虑以绿化为中心，教学区和生活区并置两侧，对应布局，同步发展。以线性骨架、弹性规划、分期实施、相对完整、逐步成型的实施模式来实现总体规划设计。

The new campus covers a total site area of 101 hectares and is constructed in three phases. It consists of the College of Fine Arts, College of Liberal Arts, College of Foreign Language, College of International Business & Management, College of Social Science, College of Science, School of Computer, College of Law, School of Intellectual Property Rights and Film & Television Production Center, etc. The master plan and architectural design aims to provide a central landscaping area flanked by the teaching area and living area which responds to each other. The Master Plan is finalized through a series specialized process including the linear skeleton analysis, flexible planning, phased implementation schedule, comparable integrity and progressive completion.

建 设 单 位：上海大学
功 能 用 途：教学
设计 / 竣工年份：1997 年 /1999 年
建 设 地 点：上海宝山区
总 建 筑 面 积 ：360000m²
总建筑高度 / 层数：65m/15 层
结 构 形 式：框架、框剪结构
合作设计单位：浙江大学建筑设计研究院

Client: Shanghai University
Function: Education
Designed/Completed: 1997/1999
Location: Baoshan District, Shanghai
Total Floor Area: 360,000m²
Total Height/Floor: 65m/ 15 Floors
Type of Structure: Framed & Shear Wall Structure
In association with Zhejiang University Architectural Design & Research Institute

上海位育中学

Weiyu High School, Shanghai

上海位育中学为上海市十个跨世纪高级寄宿中学之一，位于上海徐汇区，占地面积 11hm^2，总建筑面积为 5.2 万 m^2。主要单体项目包括图书馆、实验中心、教学楼、食堂、宿舍以及后勤用房等。

Ranked among one of the 10 state-of-the art senior boarding schools in Shanghai, it occupies a total site area of 11 hectares and houses the library, laboratory, teaching buildings, cafeteria, dormitories and back-of-the-house premises, etc.

建 设 单 位：上海位育中学筹建处
功 能 用 途：教学
设计 / 竣工年份：1996 年 /1997 年
建 设 地 点：上海市徐汇区
总 建 筑 面 积 ：52000m²
总建筑高度 / 层数：20m/4 层
结 构 形 式：框架结构

Client: Task Force of Shanghai Weiyu High School
Function: Education
Designed/Completed: 1996/1997
Location: Xuhui District, Shanghai
Total Floor Area: 52,000m²
Total Height/Floor: 20m/ 4 Floors
Type of Structure: Framed Structure

浙江机电职业技术学院

Zhejiang Institute of Mechanical & Electrical Engineering, Hangzhou

该项目位于杭州市滨江区教育科研规划内。校园功能分区由南至北分别为校前区、教学区、运动区和实习区，规划分区明确。为了规整整个校园的建筑，在形成主轴线同时，布置了若干次轴线，从而形成整个校园的统一风貌。

All functions are clearly defined from the south to the north including the pre-function zone, teaching zone, sports area and training precinct. The master planning design integrates and unifies the architectural languages within the site through the main axis and a number of secondary axes.

建 设 单 位：浙江机电职业技术学院
功 能 用 途：教学
设计 / 竣工年份：1999 年 /2001 年
建 设 地 点：杭州市滨江区教育科研规划区
总 建 筑 面 积 ：84350m^2
总建筑高度 / 层数：21m/6 层
结 构 形 式：框架结构

Client: Zhejiang Institute of Mechanical & Electrical Engineering
Function: Education
Designed/Completed: 1999/2001
Location: Binjiang District, Hangzhou
Total Floor Area: 84,350m^2
Total Height/Floor: 21m/6 Floors
Type of Structure: Framed Structure

上海第二工业大学

Shanghai Second Polytechnic University, Shanghai

上海第二工业大学新校区位于上海市浦东新区曹路镇，总用地面积为26.43hm²。学校在金海路设置主入口，以实验实训楼作为主入口的对景。由主入口进入校园，通过中央花园的绿色长廊把教学楼、实验实训楼、图文信息中心等有机地联系起来，形成学校空间构图的视觉中心，从而创造出现代教育建筑端庄典雅、环境优美的文化氛围。

The new development occupies an overall site area of 26.43 hectares. Architectural layout marks the laboratory and training building as the backdrop to the main entrance which is located at Jinhai Lu . The green arcade through the garden court organizes and connects the teaching building, laboratory and training building and the audio/visual information center to produce a visual point in the spatial arrangement and achieves a graceful and inviting landscaped atmosphere within the campus.

建设单位：上海第二工业大学
功能用途：教学
设计/竣工年份：2000年/2002年
建设地点：上海浦东新区曹路镇
总建筑面积：165000m²
总建筑高度/层数：50m/13层
结构形式：框架结构

Client: Shanghai Second Polytechnic University
Function: Education
Designed/Completed: 2000/2002
Location: Caolu County, Pudong New District, Shanghai
Total Floor Area: 165,000m²
Total Height/Floor: 50m/ 13 Floors
Type of Structure: Framed Structure

上海进才中学

Jincai Senior High School, Shanghai

上海进才中学为台湾叶根林先生捐赠的30班全日制寄宿高级中学，总用地面积 10hm^2。

学校功能分区明确。规划根据基地的周围环境情况和学校建筑的功能配置，将学校的生活区、教学区、运动区按西、中、东三个区严格分开。结合学校的规模及其现代化教学的特点，在校区内设计了礼仪广场、学习广场、休闲广场，并在建筑与建筑之间，建筑和广场之间的空间设置了一些内庭院，这不仅改善了建筑环境，同时也丰富了建筑空间，并将整个组群建筑风格定位在严谨、踏实、正气、向上的氛围上。

Donated by Mr. Ye LingGen from Taiwan, this full-time boarding school covers a total site area of 10 hectares. All functions are well defined to correspond to the immediate neighborhood and address the requirement of the education and recreation facilities. Living area, teaching area and sports area are strictly limited to the west, the middle and the east. Responding to the size of the site and the modernized education demands, a ceremony square, a learning square and a leisure square is provided in the campus and articulated with inner courtyards between the building and the squares that improves the environment and livens up the campus.

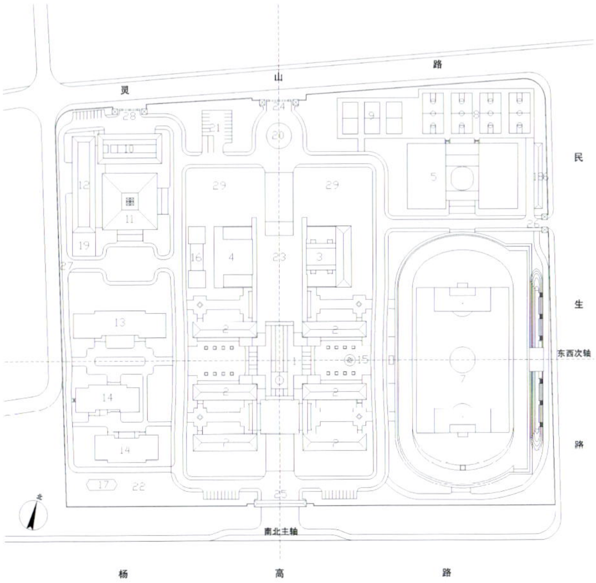

建 设 单 位：上海市教育局
功 能 用 途：教学
设计 / 竣工年份：1996 年 /1997 年
建 设 地 点：上海浦东杨高路
总 建 筑 面 积 ：50000m²
总建筑高度 / 层数：23m/5 层
结 构 形 式：框架结构
合作设计单位：台湾喻肇川建筑事务所

Client: Education Bureau of Shanghai
Function: Education
Designed/Completed: 1996/1997
Location: Yanggao Lu, Pudong, Shanghai
Total Floor Area: 50,000m²
Total Height/Floor: 23m/ 5 Floors
Type of Structure: Framed Structure
In association with C.C. YU ARCHITECT

青岛大学新校区

Qingdao University New Campus, Shandong

青岛大学新校区位于青岛市高科技开发区内，朱家洼村北面。总用地面积为21.09hm²。在保证合理利用地形和建筑最佳朝向的基础上，将教学区布置在用地东南部，生活区布置在西侧，运动区布置于东北部，学生活动中心置于西南角，办公综合楼置于主入口西侧。

设计采用“核心教学＋车流环路”的空间布置方式，使教学区做到人车分流。人们通过入口广场景观步行道，拾阶而上，最后到达校园中心广场，图文中心作为主体建筑置于轴线的终点，形成校园建筑礼仪性空间的高潮。

The new development covers an overall site area of 21.09 hectares. To maximize efficient and effective use of the site and to assure the best orientation of the building, all functions are clearly separated in different areas: the teaching buildings are designated to the south-east, living area is located at the west, sports facilities dominate the north-east, student's activity center is served at the south-west corner and the office building is situated at the west of the main entrance. The overall master plan arranges the teaching-related facilities at the center surrounded by the vehicular loop. The circulation system has been carefully manipulated to separate the pedestrian from the vehicular. The ceremonial stairs connected to the entrance plaza would permit access to the central plaza of the campus. As the main building of the campus, Library & Information Center is anchored at the terminal of the main axis.

建设单位：青岛大学
功能用途：教学
设计/竣工年份：2002年/2004年
建设地点：青岛市高科技开发区
总建筑面积：111590m²
总建筑高度/层数：22.50m/5层
结构形式：框架结构

Client: Qingdao University
Function: Education
Designed/Completed: 2002/2004
Location: Hi-tech Development Zone, Qingdao
Total Floor Area: 111,590m²
Total Height/Floor: 22.50m/ 5 Floors
Type of Structure: Framed Structure

常州信息职业学院

Changzhou College of Information Technology, Changzhou

常州信息职业学院位于常州大学城，总用地面积50.40hm²。地块的北面是校园的入口，中心广场是一个以科技、动态、人与自然为主题的大庭院。四座主要的建筑围绕中心广场的一个形象多变、自由活泼的前卫艺术品相对布置，通过几何体不规则的穿插、渗透，辅以形态各异、高低错落的水面有机构成。

Located in Changzhou University Town, the campus occupies a total site area of 50.40 hectares. The main entrance is situated at the north. The central plaza is surrounded by four main buildings and accommodates scientific, dynamic, human-oriented and natural elements that forms an ensemble in a modern and diversified architectural style to be complemented by water levels of different height, futuristic in shape.

建 设 单 位：苏中建设集团
功 能 用 途：教学
设计 / 竣工年份：2002 年 /2004 年
建 设 地 点：常州常漕路
总 建 筑 面 积 ：103820m²
总建筑高度 / 层数：34.50m/7 层
结 构 形 式：框架结构

Client: Suzhong Construction Group
Function: Education
Designed/Completed: 2002/2004
Location: Changcao Lu, Changzhou
Total Floor Area: 103,820m²
Total Height/Floor: 34.50m/ 7 Floors
Type of Structure: Framed Structure

杭州外国语学校

Hangzhou Foreign Language School, Hangzhou

杭州外国语学校新校区位于杭州市原小和山森林公园周边，规划用地26.74hm²。总体规划将校区分为教学区、生活区、运动区三大功能区块，西面的教学区综合楼群通过连廊将高中部教学楼、实验综合楼、初中部教学楼串联在一起，东面的运动区与教学区隔湖相望，绿化、水面有效地隔绝了噪音。校区东南角为学生生活区，西南面为教工生活区。

Located near Mountain Xiaohe Forest Park, the new campus covers a total site area of 26.74 hectares including the teaching zone, living zone and sports zone. The design to the teaching zone integrates the teaching building for the high school, laboratory building and teaching building for the middle school into the west site through the link building. Sports zone at the east is separated from the teaching zone by a lake and green area to minimize the noise. Living zone for students dominates the southeast corner, while for teachers and staffs is situated at the south.

建 设 单 位：杭州外国语学校
功 能 用 途：学校
设计 / 竣工年份：2002 年 / 2004 年
建 设 地 点：杭州小和山高教园区
总 建 筑 面 积：118000m²
总建筑高度 / 层数：21m/1~5 层
结 构 形 式：框架结构、网架结构、砖混结构

Client: Hangzhou Foreign Language School
Function: Education
Designed/Completed: 2002/2004
Location: Mountain Xiaohe High Education Park, Hangzhou
Total Floor Area: 118,000m²
Total Height/Floor: 21m/1～5 Floors
Type of Structure: Framed Structure, Trussed Structure, Brick & Concrete

台州中学新校区

New Campus of Taizhou High School, Taizhou

校园分为教学区、生活区、文体运动区，三个区相对独立又联系密切。通过纵横轴线和现代建筑的处理手法组织校园空间，由入口广场、中心广场、庭院、共享中厅、走廊、小桥流水共创了充满活力和生活趣味的校园空间环境。绿地通过架空层、庭院互相渗透和延伸，使校园和自然融合成一体。

The site is broken down into three major functions: teaching area, living area and sports area. They are relatively independent but also connected closely. The master plan design integrates the entrance plaza, central square, courtyard, atrium, arcades, bridges and water features in a manner that they complement each other, thus providing the most logical use of the site and achieving a lively campus ensemble. Planting elements are permeated and extended into the raised floors and the quadrangle that melds the site into the immediate natural environment.

建 设 单 位：台州中学
功 能 用 途：教学
设计 / 竣工年份：2003 年 /2005 年
建 设 地 点：临海市南湖大道
总 建 筑 面 积 ：83064 m^2
总建筑高度 / 层数：31m/7 层
结 构 形 式：框架结构

Client: Taizhou High School
Function: Education
Designed/Completed: 2003/2005
Location: South Lake Boulevard, Linhai
Total Floor Area: 83,064m^2
Total Height/Floor: 31m/ 7 Floors
Type of Structure: Framed Structure

东方绿舟——上海青少年素质教育基地

The Oriental Green Boat—Shanghai Quality-oriented Education Base for Youngsters, Shanghai

东方绿舟——上海青少年素质教育基地位于上海淀山湖畔，由知识大道区、勇敢智慧区、国防教育区、生存挑战区、科学探索区、水上运动区、运动训练区、生活实践区等八大园区组成。其中的运动训练区，分游泳馆、少儿游泳馆、篮球馆、排球馆、乒乓球馆、风雨操场等七个场馆。生活实践区由娱乐宫、商店、餐饮设施、行政办公楼、世界各地典型民居、小木屋和上海国际青少年夏令营地等组成。

The Oriental Green Boat lies in the southwest of Qingpu district in Shanghai, near the delightful Dianshan Lake. It is consisted of 8 fields: knowledge boulevard, brave wisdom field, national defense education field, survival challenge field, science exploration field, water sports field, physical training field and life experience field. The physical training field accommodates natatorium, children' s indoor swimming pool, basketball court, volleyball court, table tennis court and weathered playground. Life experience field consists of entertainment palace, retail, restaurants, administrative office, world typical vernacular building, cottage and Shanghai International Youngster' s Summer Camp.

建设单位：上海青少年素质教育基地筹建处
功能用途：素质教育
设计/竣工年份：2000年/2001年
建设地点：上海市青浦县淀山湖
总建筑面积：50000m²
总建筑高度/层数：18m/3层
结构形式：框架、网架结构

Client: Task Force of Shanghai Quality-oriented Education Base for Youngsters
Function: Quality-oriented Education
Designed/Completed: 2000/2001
Location: Dianshan Lake, Qingpu County, Shanghai
Total Floor Area: 50,000m²
Total Height/Floor: 18m/ 3 Floors
Type of Structure: Framed Structure, Trussed Structure

浙江省图书馆

Zhejiang Library, Hangzhou

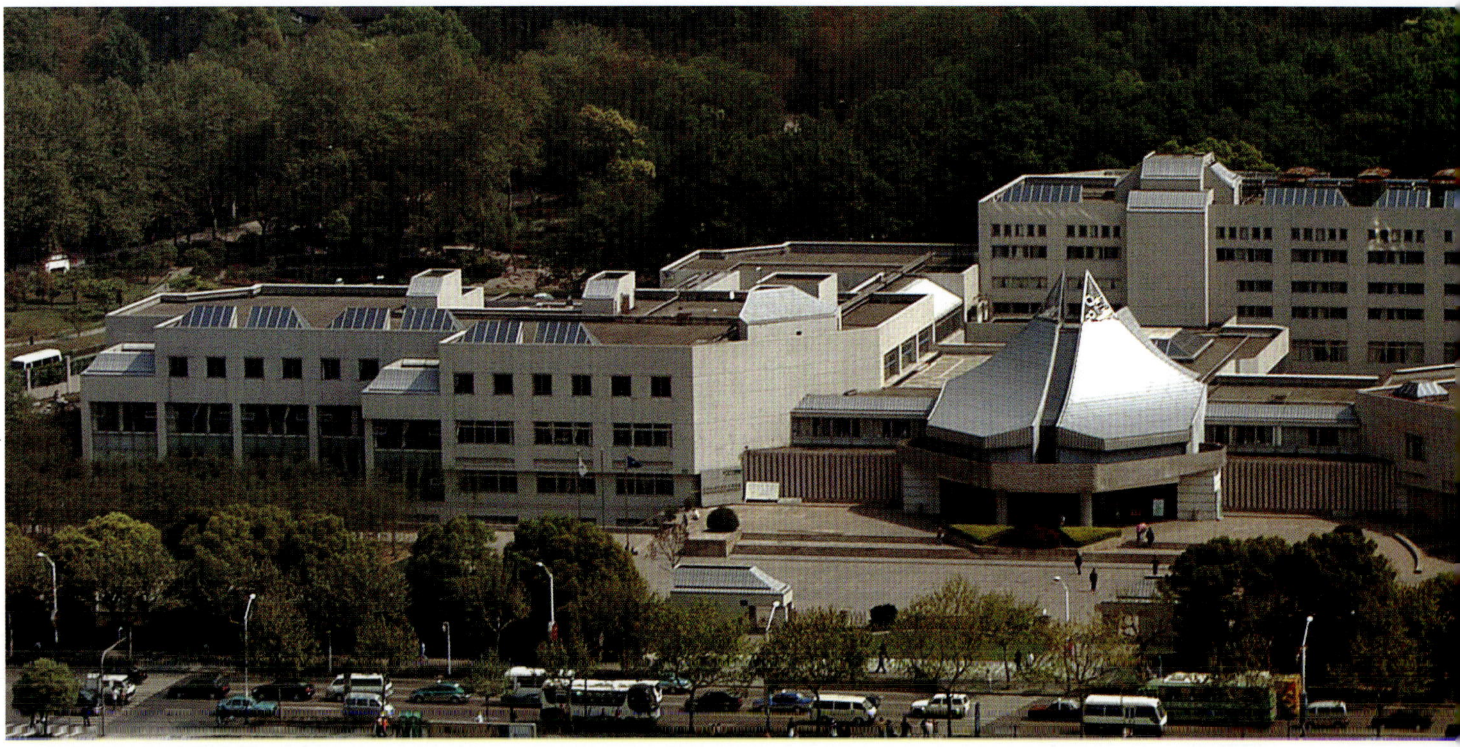

浙江省图书馆位于杭州市宝石山北麓，与西湖仅一山之隔，总平面环境设计使建筑形成低散绵延与山体充分融合的群落，既表现江南园林的通透和民居建构的亲切感，又在中轴线的建筑组合中以内外如一的手法表现出传统标志性建筑的文化纪念气息。主入口两翼用 60 片石刻书简以古朴的建筑语言展示了“书”的主题形象。建筑顶部用抽象提炼的金属色凹曲面造型构成，再现华夏传统建筑轮廓的神韵。用陶片、铝板和金属管材组合的屋顶建筑细部，把江南传统的瓦饰雕镂以现代风貌融汇在宝石山绿色背景之中。该工程被评为杭州市十佳形象建筑之一。

Located at the foot of Jade Mountain, the library is master planned to blend into the hilly terrain surrounding the site. The architectural composition is one of modern interpretation of the gardening of South China and recalls the inviting amenities of traditional indigenous residence, simultaneously refers to the functional requirement and cultural respect by unifying the ensemble on the central axis. Two wings at the main entrance are provided with 60 pieces of carved stone slip to interpret the essence of the “book” in simple language. The roofing of the building is designed in the form of bronze-colored concave to reappear the charm of the tradition Chinese architectural contour. The combination of pottery pieces, aluminum panels and metal pipes on the roof reinterprets the indigenous vernacular in the background of the Jade Mountain. The project is awarded the honor of Ten Landmark Architectures in Hangzhou.

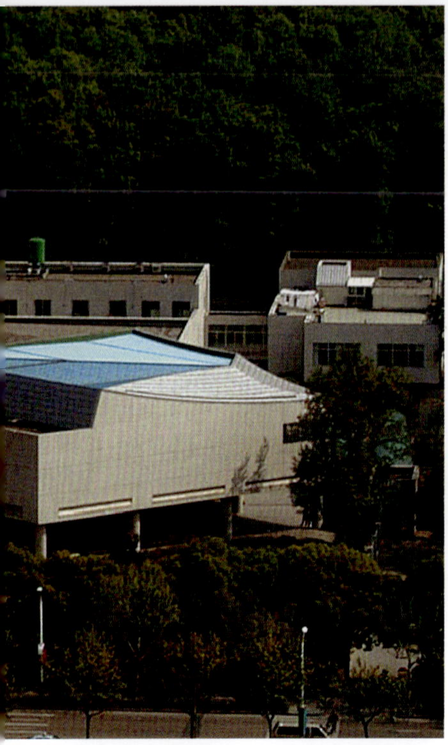

建设单位：浙江省图书馆
功能用途：大型公共图书馆
设计/竣工年份：1994年/1998年
建设地点：杭州市曙光路
总建筑面积：34000m²
总建筑高度/层数：18m/4层
结构形式：框架结构

Client: Zhejiang Library
Function: Public Library
Designed/Completed: 1994/1998
Location: Shuguang Lu, Hangzhou
Total Floor Area: 34,000m²
Total Height/Floor: 18m/ 4 Floors
Type of Structure: Framed Structure

山东省图书馆

Shandong Library, Jinan

山东省图书馆位于济南东环北路。总平面设计以庭院式的布局，采用中空的良渚玉琮为原型在建筑转角进行切割拼接，塑造了一个东方韵味的建筑轮廓。建筑的总体中空，而外形凝重端庄，既有玉琮的古朴典雅，又表现了雄浑阳刚的齐鲁风采。建筑细部的造型取材自历城隋代四门塔的石檐，建筑内部空间组合也重现了四门相通的环境层次。中央大厅设计采用7.5m的模数化大空间，布局通畅、灵活，通风采光良好，设计还特别为壁画安排了与建筑空间共生的环境，室内空间的文化气息十分浓厚。

The plan of this quadrangle-typed library draws inspiration from a special native hollow jadeware called Cong that shows an Oriental architecture. The design of void in the volume and its imposing form is reminiscent of the unsophisticated elegance of jadeware and the powerful indigenous vernacular, most importantly symbolizes the significance of its functions. The detail of roof is designed recalling of the stone eaves of pagoda in Sui Dynasty. Interior design also reinterprets the architectural composition of the historical pagoda which features four arches at four sides to be opened to each other. Central lobby in 7.5-meter modular excels with the simple orientation and efficient transport links for visitors, with high flexibility of all facilities, with effective ventilated and well-lighted interior.

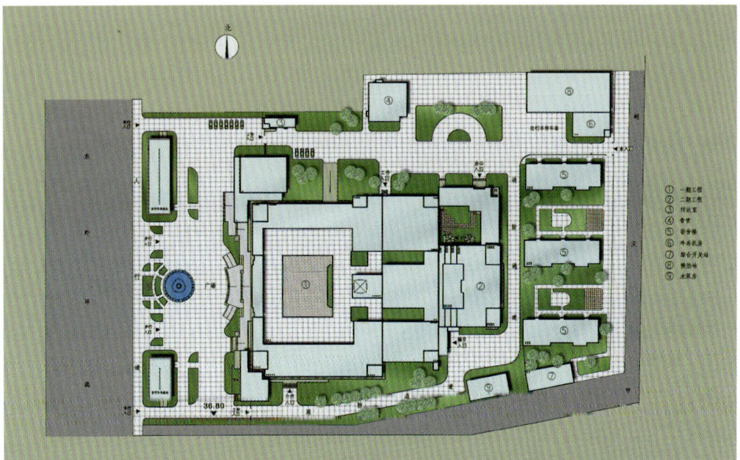

总平面图

建 设 单 位：山东省图书馆
功 能 用 途：大型公共图书馆
设计 / 竣工年份：1994 年 /2002 年
建 设 地 点：山东济南东环北路
总 建 筑 面 积 ：50000m²
总建筑高度 / 层数：42m/10 层
结 构 形 式：框架结构

Client: Shandong Library
Function: Public Library
Designed/Completed: 1994/2002
Location: Dong Huan Bei Lu, Jinan, Shandong
Total Floor Area: 50,000m²
Total Height/Floor: 42m/ 10 Floors
Type of Structure: Framed Structure

绍兴市图书馆

Shaoxing Library, Shaoxing

绍兴市图书馆位于市区延安路，北近鲁迅故居，西邻沈园和绍兴博物馆。建筑布局形成架空的庭院绿化环境，前庭广场植有绿树，曲池穿越其间，形成了建筑与环境交融的引导空间。大厅采光顶用悬挂的几何体形成富有纵深感的空间，也调节了入射的光线。

平面布局采用7.5m模数化的大空间阅览室，专设儿童阅览室、展厅、报告厅和专家研究室等。建筑外观的设计吸取了绍兴春秋时代“使乐铜房”的造型元素，使建筑呈现丰富的轮廓线。其独特的造型，使观者在感知绍兴悠久历史文明的同时，又真实地看到了一座现代的图书馆建筑。

Shaoxing Library is located at Yan'an Lu neighboring Lu Xun's Former Residence to the north and Shen Garden and Shaoxing Museum to the west. Inclusion of tree and water elements before the library makes the building blend into the natural landscaping. The clerestory roof in the lobby is provided with suspended element of geometric pattern that acknowledges spatial depth and adjusts the natural lighting from outside.

The plan layout is designed in 7.5-meter modular reading room. Spaces are dedicated to children's reading room, exhibition hall, lecture hall and specialist's research room. The architectural form is evolved from the unearthed bronze ware made in Spring-autumn Dynasty called "Shi Yue Tong Fang" which gives the project a memorable profile. Visitors would learn the historical cultural enrichment of Shaoxing city from its expressive form but also be impressed by its state-out-of-the art facilities.

建 设 单 位：绍兴市图书馆
功 能 用 途：公共图书馆
设计/竣工年份：1997 年/1999 年
建 设 地 点：绍兴市延安路
总 建 筑 面 积 ：15000m²
总建筑高度/层数：23m/5 层
结 构 形 式：框架结构

Client: Shaoxing Library
Function: Library
Designed/Completed: 1997/1999
Location: Yan' an Lu, Shaoxing
Total Floor Area: 15,000m²
Total Height/Floor: 23m/ 5 Floors
Type of Structure: Framed Structure

浙江镇海口海防历史纪念馆

Coast Defense Historic Memorial Zhenhai, Ningbo

浙江镇海口海防历史纪念馆位于镇海招宝山南麓。总占地面积17432m^2。主体建筑位于公园的东部，周围场地满铺白色细碎砾石。浮雕墙面横贯建筑。插于纪念馆西侧场地中，象征杀敌力量的纪念碑和馆体本身雄奇壮阔的造型共同烘托出凝重悲壮的纪念性气氛。

建筑造型依招宝山之势，借后海塘之雄，以西向长达几十米雄阔有力的斜面粗重石墙为设计出发点，充分体现镇海军民抗击外来侵略铜墙铁壁般的英雄气概。

The development covers an overall site area of 17,432 square meters. The main building is located at the east of the park. Sidewalks around the building are white-colored gravel cobbles pavers. A cameo wall through the building will be sensitively sculptured. The memorial environment within the site is enhanced by the magnificent monument and the majestic volume of the memorial. The clear shallow water level and the smooth sidewalks will lead the visitors to a serene and tranquil atmosphere beyond the noise park.

Strategically located adjacent to Mt. Zhaobao and depending the seawall of Zhenhai, the building with its powerful form symbolizes the heroic spirit of the local people against the foreign invasion and assures a memorable experience to the visitors.

建 设 单 位：海防历史纪念馆筹建处
功 能 用 途：纪念馆
设计 / 竣工年份：1995 年 /1997 年
建 设 地 点：镇海招宝山南麓
总 建 筑 面 积 ：3280m²
总建筑高度 / 层数：16m/2 层
结 构 形 式：框架结构
合作设计单位：东南大学建筑设计所

Client: Task Force of Coast Defense Historic Memorial
Function: Memorial
Designed/Completed: 1995/1997
Location: Mt. Zhaobao（South), Zhenhai
Total Floor Area: 3,280m²
Total Height/Floor: 16m/ 2 Floors
Type of Structure: Framed Structure
In association with Architectural Design Studio of Southeastern University

嘉兴大剧院

Jiaxing Theatre, Jiaxing

嘉兴大剧院地处嘉兴南湖新区，是一座拥有1400座的现代化多功能的综合剧场，总建面积24000m^2。设计运用内外水景的流动与融合来诠释江南水乡的风情。柱廊、星光厅、星光墙和丰富的屋面形态构成了形象鲜明的艺术宫殿。

This modern multi-functional theatre accommodates a seating capacity of 1400. Through the combination of the external and internal water features, the architectural concept is one of reinterpretation of the traditional architecture in Southern China. Portico, Star lobby, Star wall and varied roof level are designed to enhance the identity of the artistic palace.

建 设 单 位：嘉兴市文化局
功 能 用 途：演出、影视、群艺
设计 / 竣工年份：2001 年 /2003 年
建 设 地 点：嘉兴南湖新区
总 建 筑 面 积 ：24000m²
总建筑高度 / 层数：32m/3 层
结 构 形 式：框剪结构

Client: Jiaxing Cultural Bureau
Function: Performance, Cinema
Designed/Completed: 2001/2003
Location: South Lake New District, Jiaxing
Total Floor Area: 24,000m²
Total Height/Floor: 32m/ 3 Floors
Type of Structure: Framed Shear Wall Structure

杭州灵隐永福寺

Yongfu Temple, Lingyin, Hangzhou

永福寺位于杭州市灵隐景区的北端，东邻灵隐寺，是一处已荒废百年的千年古寺。

永福寺占地6.2hm^2，总建筑面积为4500m^2。根据用地特征和以展示佛教文化为主的功能要求，规划采用以现状山路为导向，散点建园的格局层层展开，建筑与自然景观融为一体。建筑单体采用江南寺院传统风格，钢筋混凝土结构，可谓推陈出新，古为今用。

Located at the north end of Lingyin Scenic Resort and adjoining Lingyin Temple at the east, the relic of this desolated ancient temple has a history of a million year and covers a site area of 6.20 hectares. The intent of the design is to interpret the Buddhism culture to the visitors. The existing mountain roads within the site are meticulously outlined and designed to lead the visitors to the destination. Different halls, designed in the style of traditional vernacular temple of Southern China, are scattered and spread over the site that melding into the terrain of the site.

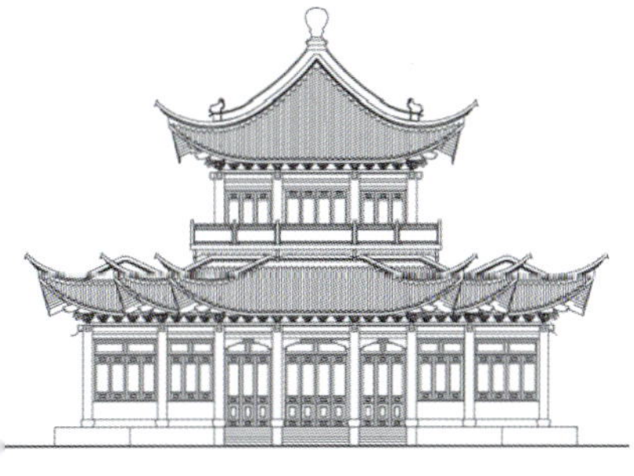

建 设 单 位：杭州市佛教协会
功 能 用 途：宗教
设计 / 竣工年份：2002 年 /2004 年
建 设 地 点：杭州市灵隐景区北端
总 建 筑 面 积 ：4500m^2
总建筑高度 / 层数：17m/2 层
结 构 形 式：钢筋混凝土仿古结构

Client: Hangzhou Buddhism Association
Function: Buddhism Culture
Designed/Completed: 2002/2004
Location: North of Lingyin Scenic Resort, Hangzhou
Total Floor Area: 4,500m^2
Total Height/Floor: 17m/2 Floors
Type of Structure: RC Structure

杭州剧院改扩建

Hangzhou Theater Renovation, Hangzhou

建设单位：浙江省文化厅
功能用途：演出、办公、宾馆、休闲、娱乐
设计/竣工年份：2000年/2003年
建设地点：杭州武林广场
总建筑面积：50000m²
总建筑高度/层数：98m/23层
结构形式：框架、框剪结构

Client: Zhejiang Province Cultural Bureau
Function: Performance, Office, Hotel, Entertainment
Designed/Completed: 2000/2003
Location: Wulin Square, Hangzhou
Total Floor Area: 50,000m²
Total Height/Floor: 98m/ 23 Floors
Type of Structure: Framed & Framed Shear Wall Structure

杭州剧院改扩建是在保留原杭州剧院观众厅、舞台主体结构的基础上，对后台部分重建。建筑立面造型根据新增功能和变化了的武林广场空间环境，设计出与其相匹配的杭州剧院新形态、新体量。挺拔高耸金属列柱衬托着凌空出挑的金属大檐口，列柱内侧的大片透明玻璃幕墙及其展现出的内部公共活动空间，使得建筑端庄宏伟，气势非凡又晶莹剔透。

The concept for this redevelopment aims to preserve the former audience level and the main structure of the stage and renovate the back of stage functions. The formal and imposing facades of the theatre reveals its design that corresponds to its new functions and the redeveloped Wulin Square nearby. The powerful metal-clad colonnades are erected to complement the grand cantilever metal eaves. A large translucent glass curtain wall behind the colonnades is provided to exhibit the interior functions.

嘉兴市图书馆

Jiaxing Library, Jiaxing

建 设 单 位：嘉兴市图书馆
功 能 用 途：图书馆
设计 / 竣工年份：2001 年 /2003 年
建 设 地 点：嘉兴市区南部
总 建 筑 面 积 ：15000m^2
总建筑高度 / 层数：24m/4 层
结 构 形 式：框架结构

Client: Jiaxing Library
Function: Library
Designed/Completed: 2001/2003
Location: Jiaxing（South）, Zhejiang
Total Floor Area: 15,000m^2
Total Height/Floor: 24m/4 Floors
Type of Structure: Framed Structure

该工程位于嘉兴市区南部，设计采用“三统一”原则，即统一柱网、统一层高、统一荷载，将各功能区集中布置于一体。南侧为阅览区，东北侧为服务区，西北侧为备餐和食堂，居中的门厅将各功能区联系在一起。一层为期刊及社科阅览、少儿阅览、备餐、餐厅、展厅以及采编等；二层为参考书阅览、外文阅览、网上阅览、视听室、自修室及报告厅；三层为古籍、自修、网络中心及内部办公。

The project is designed with the principles of unified coordinates, unified floor heights and unified loads. All functions are centralized into one building including the reading room at the south, service area at the northeast, pantry and dinning room at the northwest which are connected by the lobby in the middle. First floor is dominated by the reading room of periodicals and social science magazines, reading room for the children, pantry and canteen, exhibition hall, collection and compilation. Second floor contains the reading room of reference books and foreign books, Internet computer room, audio-visual room, study room and lecture hall. Third floor accommodates the ancient books, study room, Internet center and administrative offices.

医院建筑
Medical Care Building

台州市中心医院

Taizhou Central Hospital, Zhejiang

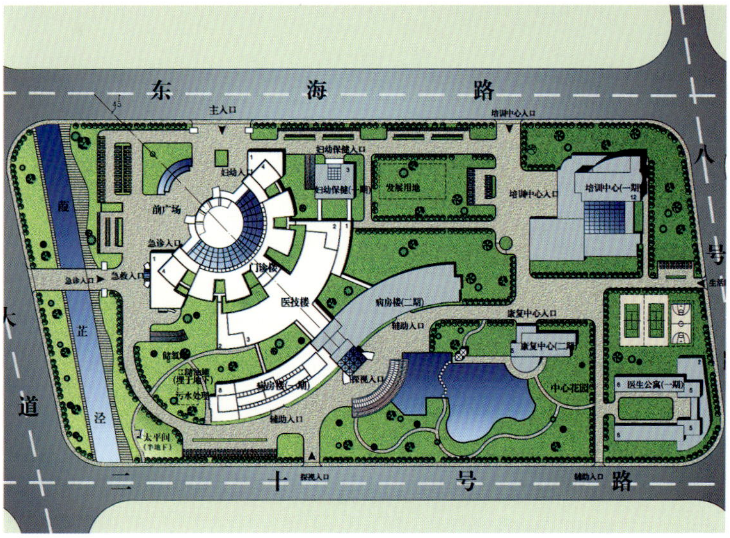

医院位于台州市经济开发区内，是一座新建的集医疗、教学、科研、预防为一体的并具专科特色的三级甲等综合性医院。总规划占地 10.73hm²，设床位 800 张，日门诊量 2400 人次。医院总体布局构思新颖独特，通过医疗中心对基地 45° 的弧形“软划分”，形成入口前广场和开放式中心花园。医疗中心布局采用医疗街连接门诊楼、医技楼、病房楼，形成有众多敞开式庭院相嵌的蝶形花园式群体建筑。门诊楼采用以玻璃顶中庭大厅为核心，医技楼连接门诊楼与病房楼。病房楼为八层正反互补弧形形体。整个建筑群体错落有致、功能布局合理、洁污流线清楚，创造出了一个现代、高雅的花园式医院。

Located in Taizhou Economic Development Zone, the design for this sophisticated hospital brings the medical treatment, research, educational programs and disease prevention services under one roof. It covers a total site area of 10.73 hectares accommodating 800 beds and receiving 2400 daily patient visits. The architecture and planning of the hospital are unique in the concept and skillful in their execution. A concourse before the entrance and an open central garden are created through cutting the site at 45 degrees by the medical treatment center. In the overall plan layout, medical street is provided to link the outpatient clinics, diagnostic and treatment functions and inpatient building which features a butterfly-shaped configuration, inserted by numbers of open courtyards. Skylighted atrium serves as the hub in the outpatient clinics. The diagnostic and treatment functions connect the outpatient clinics and inpatient building. The inpatient building is designed as a complementary front-and-back circular structure. The well-defined ensemble is staggered in height and provided with the circulation system that has been carefully designed to separate the clean from the soiled.

建设单位：台州市中心医院
功能用途：医疗
设计/竣工年份：1998年/2000年
建设地点：台州市椒江区
总建筑面积：113000m²
总建筑高度/层数：33.7m/8层
结构形式：框架结构

Client: Taizhou Central Hospital
Function: Medical Care
Designed/Completed: 1998/2000
Location: Jiaojiang District, Taizhou
Total Floor Area: 113,000m²
Total Height/Floor: 33.70m/ 8 Floors
Type of Structure: Framed Structure

玉环县人民医院

Yuhuan County People' s Hospital, Zhejiang

玉环县人民医院位于浙江玉环县珠港镇。占地7.13hm²，总投资1.35亿元，规模为400床位，1200人次/日门诊量。医院设计在总体规划、功能布局、建筑外观及环境设施上都遵循"以人为本，以病人为中心"的指导思想。宽敞的门诊大堂采用高井字梁采光顶棚以及玻璃砖采光窗，使得大堂明亮且具有柔和的光线。室内医疗主通道连接门急诊、医技、病房等部门，有很好的导向性，联系畅通，便捷。跃层天井绿化以及屋面绿化使医院环境锦上添花，也使病人、医务人员多了一些呼吸新鲜空气的好去处。

The hospital occupies a total site area of 7.13 hectares accommodating 400 beds and receiving 1200 patients per day. It costs approximately 135,000,000 RMB Yuan. The human-oriented philosophy is deeply rooted in the design of the master plan, interior, exterior and landscaping elements which emphasizes the value of the patient throughout the healing process. Ample soft sunlight is brought into the lobby of the outpatient clinics through a lighted ceiling supported by the deep-trussed beams and a glazed clerestory. Main corridor within the hospital connects to the outpatient clinics, emergency department, diagnostic and treatment functions and inpatient rooms which shows an easy and simple wayfinding system. The planted courtyard and the rooftop garden provide a welcoming amenity to both the staff and the patients.

建设单位：玉环县人民医院
功能用途：医疗
设计/竣工年份：1998年/2002年
建设地点：浙江玉环县珠港镇
总建筑面积：45500m²
总建筑高度/层数：35.8m/7层
结构形式：框架结构

Client: Yuhuan People's Hospital
Function: Medical Care
Designed/Completed: 1998 / 2002
Location: Zhugang Town, Yuhuan County, Zhejiang
Total Floor Area: 45,500m²
Total Height/Floor: 35.80m/ 7 Floors
Type of Structure: Framed Structure

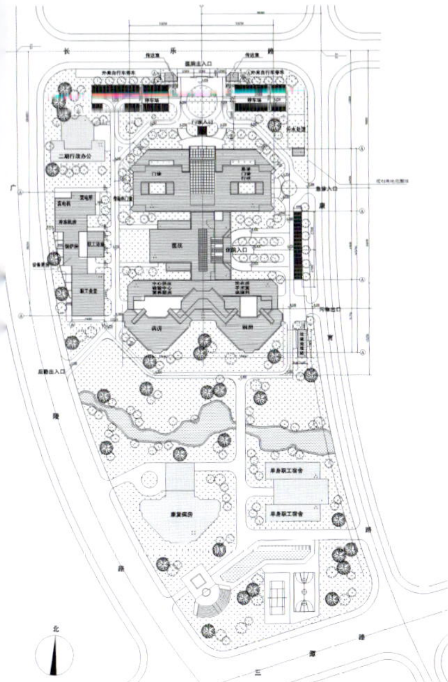

总平面布置图
1:500

上海仁济医院（一期）

Renji Hospital（Phase I），Shanghai

上海仁济医院东院是上海第二医科大学附属的一所综合性教学医院。医院位于上海市浦东新区，占地约$4.8hm^2$，规模为500床位，2000人次/日门诊量，总投资2.5亿元。设计从医院建筑的性质、功能的特征以及创造良好的环境着手，总体分区自东向西分为门急诊区、病房区、附属用房区等。主体低层部分均设有庭院和中庭绿化，交通流线合理、人车有序、洁污分流。建筑外观造型简洁、明快、层次分明、错落有致。室内空间流畅，行医就医流程便捷，自然通风采光良好，设计还注重创造了良好的绿化环境及现代建筑气息。

As an affiliated hospital to the Shanghai Second Medical University, this comprehensive teaching hospital covers a total site area of 4.80 hectares accommodating 500 beds and receiving 2000 daily patient visits. The development costs approximately 250 million RMB Yuan. The concept of the design is focused to facilitate comprehensive state-of-the art medical services within a sunlit humane environment that responds the surrounding architecture and immediate environment. From the east to the west the site is divided into outpatient clinics, emergency department, inpatient nursing and support spaces, etc. All lower levels in the main structure are provided with gardens and plants. The circulation system has been carefully designed to provide good communication between departments and proper separation of traffic types. Separate entrances and proper design of the interior circulation systems are provided for the separation of inpatients from outpatients, clean from soiled, and other similar requirements. The appearance of the building is featured sophisticated simplicity.

建 设 单 位：上海第二医科大学附属仁济医院
功 能 用 途：医疗
设计 / 竣工年份：1994 年 /1998 年
建 设 地 点：上海市浦东新区
总 建 筑 面 积：41200m^2
总建筑高度 / 层数：65m/16 层
结 构 形 式：框剪结构

Client: Shanghai Second Medical University Affiliated Renji Hospital
Function: Medical Care
Designed/Completed: 1994/1998
Location: Pudong New District, Shanghai
Total Floor Area: 41,200m^2
Total Height/Floor: 65m/ 16 Floors
Type of Structure: Framed Shear Wall Structure

杭州邵逸夫医院（一期）

Sir Run Run Shaw Hospital （Phase I），Hangzhou

建 设 单 位：杭州邵逸夫医院
功 能 用 途：医疗
设计 / 竣工年份：1990 年 /1992 年
建 设 地 点：杭州市庆春东路 3 号
总 建 筑 面 积 ：36000m^2
总建筑高度 / 层数：64.6m/16 层
结 构 形 式：框剪结构

Client: Sir Run Run Shaw Hospital, Hangzhou
Function: Medical Care
Designed/Completed: 1990/1992
Location: No. 3 Qingchun Lu (East), Hangzhou
Total Floor Area: 36,000m^2
Total Height/Floor: 64.60m/ 16 Floors
Type of Structure: Framed Shear Wall Structure

邵逸夫医院位于杭州市秋涛路与庆春路交叉口。占地 5.3hm^2，一期床位 400 床，标准层为 40 床护理单元。该医院系浙江省人民政府出资，香港著名爱国人士邵逸夫先生赞助，美国罗马琳达大学医学院参加管理的浙江医科大学附属教学医院。医疗重点为肿瘤及心血管系统疾病，门诊科室包括除妇产科、儿科、传染病科（已有专科教学医院）以外的其他各科室。

Sir Run Run Shaw Hospital is a comprehensive state-of-the art medical service covering an overall site area of 5.3 hectares and accommodating 400 beds with 40-bed nursing units on typical levels. It is funded by Zhejiang Provincial Government, sponsored by the well-known patriots Sir Run Run Shaw from Hong Kong and managed by Loma Linda University Medical Center. It is also an affiliated teaching hospital to Zhejiang Medical University. The hub of the medical care services consists of the tumor clinic and the cardiovascular clinic. The outpatient clinics include all the medical departments except the OB/GYN clinic, Pediatrics and infectious clinic.

金华儿童医院

Jinhua Children's Hospital, Zhejiang

建 设 单 位：横店集团
功 能 用 途：医疗
设计 / 竣工年份：2002 年 /2003 年
建 设 地 点：金华市东莱路
总 建 筑 面 积：37000m²
总建筑高度 / 层数：36.8m/9 层
结 构 形 式：框架结构

Client: Hengdian Group
Function: Medical Care
Designed/Completed: 2002/2003
Location: Donglai Lu, Jinhua
Total Floor Area: 37,000m²
Total Height/Floor: 36.80m/ 9 Floors
Type of Structure: Framed Structure

根据儿童医院的特点，设计将所有门诊诊室、特检科、急诊均布置在底层，以方便小患者就诊，二层布置手术室、实验室、教室及办公用房，各医疗、医技科室联系紧密。手术室设计采用先进的二通道方式，同时有ICU、NICU、CCU等。住院部每层有两个护理单元，充分利用楼、电梯，节约交通面积。

Respectful of the need of the children, all the OPD clinics, special diagnostic functions and emergency clinic are housed in the ground floor to facilitate the children. Second floor is served by surgery, laboratory, classrooms and offices. The design provides a good communication between all departments and clinics. Surgery is provided with state-of-the art 2-circulaiton systems and ICU, NICU and CCU. Two nursing units are located on each floor in the inpatient building to maximize the use of elevators and the traffic areas.

浙江省台州医院

Zhejiang Taizhou Hospital, Zhejiang

建 设 单 位：浙江省台州医院
功 能 用 途：医疗
设计 / 竣工年份：1996 年 /1999 年
建 设 地 点：浙江省临海市老城区
总 建 筑 面 积 ：30619m²
总建筑高度 / 层数：76m/21 层
结 构 形 式：框剪结构

Client: Zhejiang Taizhou Hospital
Function: Medical Care
Designed/Completed: 1996/1999
Location: Old City, Linhai, Zhejiang
Total Floor Area: 30,619m²
Total Height/Floor: 76m/ 21 Floors
Type of Structure: Framed Shear Wall Structure

浙江省台州医院地处浙江省临海市老城区，为三级甲类综合医院。设计床位 850 张，附设有监护中心、产科、手术室、中心供应等。病房大楼采用了点式高层的布置，与门诊大楼和医技楼通过连廊相互联系，形成医疗区内合理的功能分区。标准层平面采用了长方形和三角形组合，并由此形成了一个向心性的护理单元，护士站布置在中心位置。设计在大楼顶部采用收分退台等手法塑造了一个八角形的古塔形式的尖顶，以与老城区的原有建筑相呼应。

The comprehensive tertiary-level hospital accommodates 850 beds and includes ICU/CCU, obstetrical clinic, surgery and central supply, etc. The inpatient building of tower structure connects the OPD clinics and diagnostic and treatment functions through corridors that defines an organic arrangement between departments. All the standard floors are configured in the geometric pattern of the rectangle and triangle to allow for central nurse stations with nursing units around it. Roof level steps back that creates an octagonal pinnacle corresponding to the existing buildings in the old city.

台州市立医院门诊综合楼与住院楼

Outpatient & Inpatient Building Taizhou Municipal Hospital, Zhejiang

门诊楼

建设单位：台州市立医院
功能用途：医疗
设计/竣工年份：1999年/2004年
建设地点：台州市椒江区中山路
总建筑面积：18000m²(门诊楼)，37650m²(住院楼)
总建筑高度/层数：70m/ 17层，59m/ 10层
结构形式：框剪结构

Client: Taizhou Municipal Hospital
Function: Medical Care
Designed/Completed: 1999/2004
Location: Zhongshan Lu, Jiaojiang District, Taizhou
Total Floor Area: 18,000m²(OPD Clinics), 37,650m² (Inpatient Building)
Total Height/Floor: 70m/ 17 Floors, 59m/ 10 Floors
Type of Structure: Framed Shear Wall Structure

住院楼

台州市立医院位于台州市椒江区，为600床的三级甲类综合医院。门诊综合楼与住院楼、外科大楼围合形成"品"字形的中心广场，门诊综合楼为17层，建筑面积18000m²，建筑高度70.20m。门诊楼裙房内设置了两个采光中庭，功能用房围绕中庭设置，每层均设有集中的候诊区域，住院综合楼为10层，建筑面积为37650m²，高度为59m，设有住院部，手术中心、产科中心等。

The design for this comprehensive tertiary-level 600-bed hospital allows effective and efficient use of the site. The central courtyard is embraced by the OPD clinics, the inpatient building and the surgical clinics. The OPD clinics occupies 17 levels and a total floor area of 18,000 Square meters with two skylighted atriums in its podium structure which are surrounded by functional rooms and distributed by the centralized waiting rooms on each level. The inpatient building with 10 levels and a total floor area of 37,650 Square meters accommodates the inpatient, surgery and obstetrical department, etc.

浙江医院综合门诊楼

Comprehensive Outpatient Building Zhejiang Hospital, Hangzhou

建 设 单 位：浙江医院
功 能 用 途：门急诊、体检中心
设计 / 竣工年份：2004 年 / 2007 年
建 设 地 点：杭州市灵隐路 9 号
总 建 筑 面 积：17000m^2
总建筑高度 / 层数：12m/3 层，地下 2 层
结 构 形 式：框架结构

Client: Zhejiang Hospital
Function: Medical Care
Designed/Completed: 2004/2007
Location: No. 9 Lingyin Lu, Hangzhou, Zhejiang
Total Floor Area: 17,000m^2
Total Height/Floor:12.00m/ 3 Floors(above ground)+ 2 Floors(below ground)
Type of Structure: Framed Structure

浙江医院坐落在西湖风景区灵隐路上。新建 1000 人次 / 日的门诊楼，建筑采用 2～3 层的高度，分散的体量，淡雅的色彩，以尽量融入周围的环境中。主入口以弧廊与旧的保留建筑相连，2 层的高度使入口广场的尺度更为宜人。门厅处为 3 层通高的中庭空间，上部覆以玻璃顶棚。

底层中庭与医疗街一点一线，形成交通的骨架。南面四个诊疗单元与医疗街及中庭一一贯通，构成终端式的病患流线，显著提高了内部的卫生环境质量。

Located in Lingyin Lu in the Westlake Scenic Resort, this renovated and expanded complex is designed to comply with the City's planning criteria. The two or three stories buildings tinted with obscure colors, spreading over the site in harmonious with the immediate environment, would accommodate the newly-built outpatient building and receive 1000 daily patient visits. The two-storey main entry connects the preexisting building through the circular arcade and features a humane scale to the concourse. The three-story atrium in the lobby is covered with lighted ceiling. The atrium in the ground level and the medical street is served as the main circulation skeleton that passes through four diagnostic and treatment units at the south which minimizes the opportunities of cross infection.

常州市第三人民医院门急诊病房综合楼

Outpatient, Acute Care, Inpatient Building of No.3 People' s Hospital,Changzhou

建设单位：常州市第三人民医院
功能用途：医疗
设计/竣工年份：2003年/2009年
建设地点：江苏省常州市
总建筑面积：46414m²
总建筑高度/层数：84.3m/20层
结构形式：框架剪力墙结构

Client: No.3 Changzhou People's Hospital
Function: Medical Care
Designed/Completed: 2003 / 2009
Location: Changzhou, Zhejiang
Total Floor Area: 46,414m²
Total Height/Floor: 84.30m/ 20 Floors
Type of Structure: Framed Shear Wall Structure

常州市第三人民医院位于312国道与兰陵路的交汇处，交通十分便利。新建医院占地约17082m²，600床位，2000人次/日门诊量。规划将主楼与裙房一字排列，主体建筑与门急诊部形成T形，内外部交通流线方便通畅，洁污分流明确，功能完整，为病人创造了一个舒适轻松、体现人性化的就医环境。

建筑立面处理简洁大方，轻灵通透，主楼的形象高大挺拔，极富表现力，更因其内部庭院环境幽静，绿化丰富，使它成为常州市又一个崭新的医院特色的代表。

Located at the junction between No.312 State-graded Thoroughfare and Lanling Lu, the development enjoys an easy and accessible transportation system. It occupies a total site area of 17,082 square meters and accommodates 600 beds and 2000 daily patient visits. The main building is configured to align with the podium structure. The circulation system has been carefully designed to provide good communication between departments and proper separation of inpatients from outpatients, clean from soiled, and other similar requirements. The design aims to facilitate comprehensive, state-of-the art medical services within a sunlit humane environment. The expressive form of the main building which stands against the landscaping environment gives a new identity to the location.

永康市第一人民医院

No.1 Yongkang People's Hospital, Zhejiang

这是一所拥有1200床位，3600人次/日门诊量的超大型现代化综合医院。医院平面设计采用多个功能组团式布置，流线相对集中。通过门诊大厅、连廊、庭院组合空间，使阳光、绿地成为空间的主角，使人与自然相互融合、交流，为病人创造出一个良好的医疗环境，为医生、工作人员创造了一个优美的工作环境。门急诊医技楼的平面组织是以点（门诊大厅）为中心，以线（连接各门诊单元、医技楼、急救中心的交通廊）为纽带，以面（急救中心、门诊、医技等）为各功能区块共同组成一个功能齐全、联系便捷、空间丰富的医院建筑群。

This state-of-the art mega hospital accommodates 1200 beds and 3600 daily patient visits. The planning principle is based on the concept of creating a comprehensive medical service which is incorporated and grouped through the outpatient lobby, the arcades and gardens, while allowing the sunlight and plants as the main elements within the site that provides a human-oriented hospital melding into the nature and provides welcoming amenities to the patients, doctors and the staff.The organization concept to the outpatient building, acute care, diagnostic and treatment functions addresses the outpatient lobby as the center and locates the arcades to connect each outpatient clinics, diagnostic and treatment functions and the acute care department. A full-quipped and easy wayfinding configuration with spatial richness is achieved to each functional premise through the articulation of acute care department, OPD clinics and diagnostic and treatment functions, etc.

建 设 单 位：永康市第一人民医院	Client: No.1 Yongkang People's Hospital
功 能 用 途：医疗	Function: Medical Care
设计 / 竣工年份：2003 年 /2006 年	Designed/Completed: 2003 / 2006
建 设 地 点：浙江省永康市	Location: Yongkang, Zhejiang
总 建 筑 面 积 ：128370m^2	Total Floor Area: 128,370m^2
总建筑高度 / 层数：42m/11 层	Total Height/Floor: 42m/11 Floors
结 构 形 式：框架结构	Type of Structure: Framed Structure

温岭市第一人民医院

No. 1 People's Hospital Wenling, Zhejiang

建 设 单 位：温岭市第一人民医院
功 能 用 途：医疗
设计 / 竣工年份：1995 年 /2000 年
建 设 地 点：浙江温岭市
总 建 筑 面 积 ：27500m²
总建筑高度 / 层数：79.56m/19 层
结 构 形 式：框剪结构

Client: No. 1 Wenling People's Hospital
Function: Medical Care
Designed/Completed: 1995/2000
Location: Wenling, Zhejiang
Total Floor Area: 27,500m²
Total Height/Floor: 79.56m/ 19 Floors
Type of Structure: Framed Shear Wall Structure

温岭市第一人民医院地处浙江省温岭市，为三级甲类综合医院。新建的住院综合大楼采用了长方形和半圆形组合形成的平面构成，可以争取超过 90% 的南向病房和形成向心性的护理单元。岛形护士站布置在中心位置。设计床位 780 张，附设有监护中心、产科、手术中心等部门。

The development for this tertiary-graded comprehensive hospital includes the newly built inpatient building to be designed in rectangular and semi-circular forms so that provides nearly 90% patient rooms facing the south. Nurse's stations are located in the center. It accommodates 780 beds with supporting services such as ICU, obstetrical department and surgery, etc.

南通医学院附属第一医院综合病房楼

Inpatient Building of The First, Affiliated Hospital of Nantong Medical College, Jiangsu

建 设 单 位：南通医学院附属第一医院
功 能 用 途：医疗
设计/竣工年份：2001年/2004年
建 设 地 点：江苏省南通市濠河边
总 建 筑 面 积：62747m²
总建筑高度/层数：80.2m / 19层
结 构 形 式：框剪结构

Client: The First Affiliated Hospital of Nantong Medical College
Function: Medical Care
Designed/Completed: 2001/2004
Location: Nantong, Jiangsu
Total Floor Area: 62,747m²
Total Height/Floor: 80.20m/ 19 Floors
Type of Structure: Framed Shear Wall Structure

本工程位于江苏省南通市中心的濠河边，是在原有医疗用房部分拆除后新建的800床综合医院。设计思想是一切以病人为中心，方便病人治病并方便医生行医。

建筑外观造型力图体现原创性和标志性，并着重于细部的处理及整体形式的创造。西南面临濠河一侧造型顶部作退台处理，使之成为水面上两叶风帆的造型，形式活泼新颖，并突出了建筑的标志性。

Located in the downtown of the city near Hao River, this renovated and expanded comprehensive hospital accommodates 800 beds. The concept design focuses on the human-oriented principles. The design to the appearance aims to provide an icon building in the city while accentuating the textures in details and the form of the ensemble. Roof level at the southwestern side facing Hao River steps back reminiscent of two pieces of sail on the water that gives a new identity to the location.

宁波第六人民医院

No.6 People's Hospital Ningbo, Zhejiang

建 设 单 位：宁波第六人民医院
功 能 用 途：医疗
设计 / 竣工年份：2004 年 / 2008 年
建 设 地 点：宁波市东大门
总 建 筑 面 积：47370m²
总建筑高度 / 层数：90m/21 层
结 构 形 式：框剪结构

Client: No.6 Ningbo People's Hospital
Function: Medical Care
Designed/Completed: 2004 / 2008
Location: East Gate, Ningbo, Zhejiang
Total Floor Area: 47,370m²
Total Height/Floor: 90m/ 21 Floors
Type of Structure: Framed Shear Wall Structure

宁波第六人民医院位于中山东路与福明路交汇处，总占地面积为3.2hm²，2000床位，3000人次/日门诊量。该项目为新住院楼扩建，同时整合已有的医院建筑，对全院进行重新布局。

设计强调门诊中心、急诊中心、医技中心、手术中心、住院部及后勤服务中心的有机联系，使改建以后医院总体规划更趋合理。

The development covers a total site area of 3.20 hectares. It accommodates 2000 beds and 3000 daily patient visits. The organization concept aims to expand the inpatient building and renovate preexisting buildings.

The design emphasizes the organic connection between the OPD clinics, emergency department, diagnostic and treatment center, surgery, inpatient department and support functions to provide a well-defined comprehensive hospital.

仙居县人民医院

Xianju County People's Hospital, Zhejiang

建设单位：仙居县人民医院
功能用途：门急诊、医技、病房楼
设计/竣工年份：2003年/未建
建设地点：浙江省仙居县
总建筑面积：$93000m^2$
总建筑高度/层数：40m/10层
结构形式：框架结构

Client: Xianju County People's Hospital
Function: Medical Care
Designed/Completed: 2003 / unbuilt
Location: Xianju County, Zhejiang
Total Floor Area: $93,000m^2$
Total Height/Floor: 40m/10 Floors
Type of Structure: Framed Structure

仙居县人民医院一次规划，分二期建造，建筑规模为700病床，1500人次/日门诊量。一期工程包括门急诊楼、医技楼、病房楼、营养厨房，二期工程包括行政楼、康复病房楼、医生值班公寓、传染病房。总体布局将医院分为五大功能区：门急诊医疗区、住院康复区、职工生活区、后勤保障服务区、独立的传染病区。设有全天候医疗主街，将门急诊、医技和病房楼、营养厨房紧密相联，组成医疗区，以方便病患诊治。

The proposed project contains 700 beds and 1500 daily patient visits. The concept for this design would meet the phasing requirements of the program and the overall architectural layout. Phase One includes the outpatient and acute care building, diagnostic and treatment functions, inpatient building and nutrition kitchen. Phase Two contains the administration building, rehabilitation inpatient building, apartment for the doctors on duty and inpatient building for infectious clinics. In the overall plan layout, the hospital is divided into 5 areas: medical care area for outpatient and acute care clinics, inpatient rehabilitation area, living area for staff, supporting area and independent infectious disease area. It is also provided with the all-weathered medical street that integrates the outpatient and acute care clinics, diagnostic and treatment functions, inpatient building and nutrition kitchen into the medical care area that facilitates the doctors and the patients.

东阳市人民医院新病房楼

Inpatient Building, Dongyang People's Hospital, Zhejiang

建设单位：东阳市人民医院
功能用途：医疗
设计/竣工年份：2000年/2003年
建设地点：浙江省东阳市
总建筑面积：35000m^2
总建筑高度/层数：83.8m/19层
结构形式：框筒结构

Client: Dongyang People's Hospital
Function: Medical Care
Designed/Completed: 2000/2003
Location: Dongyang, Zhejiang
Total Floor Area: 35,000m^2
Total Height/Floor: 83.80m/ 19 Floors
Type of Structure: RC Framed Corewall Structure

这是一个改造项目，规模为门诊人次1500/日、500床位并达到医院能获得较好的景观环境和合理的内部交通。设计保证了每间病房朝南面对医院中心花园，并且护理路线最便捷。建筑主体特色鲜明。

This renovated project accommodates 500 beds and receives 1500 patients per day. It was the principle intention of the design to create a landscaping environment within the site and to the immediate surrounding. The interior circulation system has been carefully designed to provide good communication between departments and proper separation of traffic types. Inpatient units are designed to provide sunlight from the south to each of the patient rooms. Nurse's station located near the center allows the shortest distance from the patients.

常州市儿童医院门急诊病房综合楼

Outpatient Clinics, Acute Care Clinics, Inpatient Building, Changzhou Children's Hospital, Jiangsu

建 设 单 位：常州市儿童医院
功 能 用 途：医疗
设计 / 竣工年份：2002 年 /2005 年
建 设 地 点：江苏省常州市延陵东路
总 建 筑 面 积：20141m^2
总建筑高度 / 层数：24m/6 层
结 构 形 式：框剪结构

Client: Changzhou Children's Hospital
Function: Medical Care
Designed/Completed: 2002/2005
Location: Yanling Lu (East), Changzhou, Jiangsu
Total Floor Area: 20,141m^2
Total Height/Floor: 24m/ 6 Floors
Type of Structure: Framed Shear Wall Structure

该院位于常州市延陵东路，处于常州市历史文化保护区域的附近。总占地面积为12846m^2，300床位，1500人次／日门诊量。该楼是原老病房楼扩建的设计项目。针对本工程特殊规划环境要求，建筑立面处理上借鉴了中国园林的处理手法，坡屋顶、挑檐、门洞墙以及月亮门式的窗的运用，新、老楼之间设置连廊连接，使新老建筑有机结合。院内人车分流明确，环境优美，整个大楼造型新颖，是一座大方得体的建筑。

Located near the historical cultural region of the city, the proposed development covers a total site area of 12,846 square meters. It accommodates 300 beds and 1500 daily patient visits which shall be integrated into the preexisting building through the corridors. Respectful of the City's planning criteria to the region, the design for the facades acknowledges the essence of Chinese gardening by introducing the pitched roof, cornice, gateway wall, moon gate and window that gives the project a memorable profile. Circulation system is meticulously designed to separate the pedestrians from the vehicles.

住宅建筑

Residential Building

杭州颐景园小区

Yijingyuan Housing Development, Hangzhou

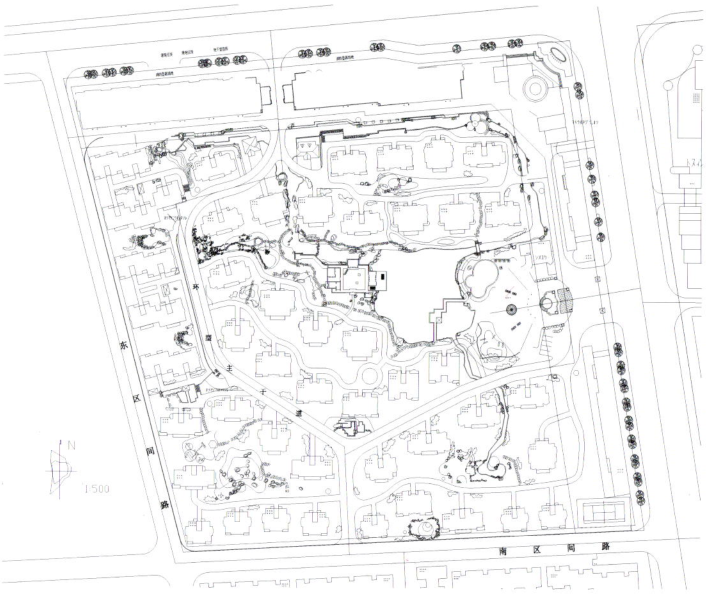

颐景园小区位于杭州市文二西路与杭三路交汇地段，总用地面积为7.95hm^2。小区总体环境设计优美。在小区的中心地带，布置了大面积的人工湖水系，围绕人工湖又设置了具有我国江南地方特色的园林小品，为住户提供了诗情画意的居住环境。

The housing covers a total site area of 7.95 hectares. Great man-made lake is located in the center of the site surrounded by Chinese gardening ingredients to promote a lively and poetic living quarter to the residents.

建设单位：杭州三盛房产有限公司
功能用途：住宅小区
设计/竣工年份：1998年/2000年
建设地点：杭州市文二西路、杭三路交叉口
总建筑面积：120800m^2
总建筑高度/层数：18m/6层
结构形式：砖混结构

Client: Hangzhou Sansheng Real Estate Co., Ltd.
Function: Housing
Designed/Completed: 1998/2000
Location: Wen' er Lu（West), Hangsan Lu, Hangzhou
Total Floor Area: 120,800m^2
Total Height/Floor: 18m/ 6 Floors
Type of Structure: Concrete & Brick Structure

苏州香城花园（一期）

Fragrant Town (Phase Ⅰ), Suzhou

工程位于苏州城北相城区南部，紧邻沪宁高速公路与苏嘉杭高速公路，南面为大面积的生态绿化公园，交通与环境条件优越。相城区是苏州市按花园城市规划建设的新城之一，随着住宅建设的蓬勃发展城市配套日益完善，居住的氛围和生活便利条件正迅速提升。为了把人居的舒适、交通的便利和景观的共享相结合，设计以“田”字形的结构进行总平面布局，用环型通道组织机动车交通，联系小区各出入口和地下车库入口；中央部分“十”字形休闲景观带将小区自然划分为四个组团，并组织人行散步系统。高层住宅主要布置在北面，沿南北向景观带只摆放点式小高层建筑，保证了小区采光、通风与视觉的通透。小区内同时设计了配套幼儿园和会所，考虑到和二、三期工程的衔接，均邻城市道路布置。建筑造型采用现代建筑手法，体量错落丰富，造型简洁明快，强调居住建筑的亲和力。

Situated near Shanghai-Ningbo Highway and Suzhou-Jiaxing-Hangzhou Highway, the site is bounded the grand ecologic landscaping park at the south and enjoys splendid location and accessible traffic. Xiangcheng district is a new developed urban center that is constructed to the criteria of garden city. It is expanded and improved with the prosperous development of housing market. In order to integrate the living comfort, traffic convenience and share of landscaping elements, the overall site planning arrangement is designed in 田(one of Chinese character)shape. Looped road is designed to the vehicular circulation and to connect each access to the site and the covered parking. Four clusters within the site are divided and articulated by the central crisscross-shaped leisure and planted garden which is provided with sidewalks. High-rise buildings are mainly located at the north. The medium-rise tower buildings are distributed along the north-southern planted area that assures lighting, ventilation and undisturbed view to the site. The supporting functions include the kindergarten and clubhouse that are located along the city's street level considering the connection with the Phase II and Phase III. The architectural building is designed in a modern language with "terraced"structure and simple form to enforce the human scale of housing building.

建 设 单 位：苏州海发置业有限公司
功 能 用 途：住宅及配套公建
设计 / 竣工年份：2003 年 /2006 年
建 设 地 点：苏州市相城区日益路 / 铁西路
总 建 筑 面 积 ：325000m²
总建筑高度 / 层数：52.2m/18 层
结 构 形 式：框剪结构

Client: Suzhou High Far Real Estate Co., Ltd.
Function: Housing, Supporting Functions
Designed/Completed: 2003/2006
Location: Riyi Lu/Tiexi Lu, Xiangcheng District, Suzhou
Total Floor Area: 325,000m²
Total Height/Floor: 52.20m/ 18 Floors
Type of Structure: Framed Shear Wall Structure

黄山广宇 · 江南新城

Southern China New Town • Mt. Huangshan, Anhui

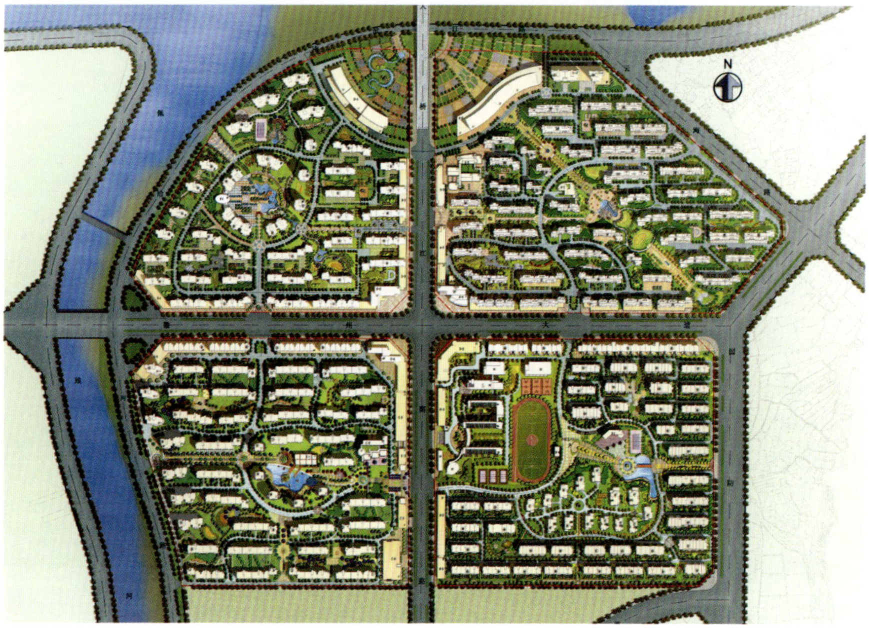

黄山广宇·江南新城位于黄山市中心城区新安江南岸，项目东望柏山，南接徽杭高速公路入口，西依佩浪河，北临新安江，拥有优越的人文和地理环境。江南新城占地约 66.7hm^2，总建筑面积约 80 万 m^2，是黄山市超大规模的高品质居住社区。

江南新城项目按用地划分为A、B、C、D四个独立的区块，A地块主要为多层住宅，总建筑面积约21万m^2，B地块开发排屋、叠层和别墅，总建筑面积约 12 万 m^2，C地块为多层住宅，结合高层住宅，总建筑面积约21万m^2，项目还包括一所 36 班九年一贯制学校，两所幼儿园、宾馆、写字楼、农贸市场等各项公用设施，共同组成整个居住区的配套和功能的延伸。

Located at the southern bank of Xin'an River, the development is bounded Mt. Baishan at the east, Anhui-Hangzhou Express at the south, Peilang River at the west and Xin' an River at the north. It covers a site area of 66.70 hectares and designed as a high quality mega residential community in Huangshan.

The site is divided into four independent lots. Lot A contains 210,000-square-meter multi-storey housing. Lot B consists of 120,000-square-meters town house, low-storey residence and villa. Lot C is taken up by 210,000-sqaure-meters multi-storey and high-rise housing. In addition, the project is provided with a 36-class nine-year compulsory education school, two kindergartens, hotel, office building and farm product market as the supporting functions.

建设单位：黄山广宇·江南新城
功能用途：住宅
设计/竣工年份：2003年/在建
建设地点：安徽黄山市中心城区
总建筑面积：800000m²
结构形式：框架、砖混

Client: Huangshan Cosmos Group Co., Ltd.
Function: Housing
Designed/Completed: 2003/under construction
Location: Mt. Huangshan, Anhui
Total Floor Area: 800,000m²
Type of Structure: Framed Structure, Concrete & Brick

上海秋月枫舍小区

"Autumn Moon & Maple" Housing Development, Shanghai

秋月枫舍小区占地面积为2.92hm²。小区由7幢高层，1幢3层联体公寓及东西两侧的商业裙房围合而成。平面布局主次分明，错落有致，活泼生动，富有特色。该小区已获2000年建设部"创新风暴"全国住宅设计夺标活动组委会特别金奖。获2001年建设部"新世纪人居经典住宅小区方案竞赛"综合大奖。

The development occupies an overall site area of 2.92 hectares. It is composed of 7 high-rise buildings and one 3-storey attached apartment building flanked by a commercial podium structure to the west and east sides. The well-proportioned building is staggered in height to create a distinguished identity. The development has been awarded the honor of Special Medal by the Organizing Committee of "Innovation Storm" National Residential Design Competition in 2000 and the prize of New Century Housing Community Concept Design Competition 2001 by the Ministry of Construction.

建设单位：上海沙田房地产开发有限公司
功能用途：住宅
设计/竣工年份：2000年/2001年
建设地点：上海市镇坪路
总建筑面积：84030m^2
总建筑高度/层数：56m～78m/18～28层
结构形式：框剪结构

Client: Shanghai Shatian Real Estate Development Co. Ltd.
Function: Housing
Designed/Completed: 2000/2001
Location: Zhenping Lu, Shanghai
Total Floor Area: 84,030m^2
Total Height/Floor: 56～78m/ 18～28 Floors
Type of Structure: Framed Shear Wall Structure

上海新湖明珠城

Xinhu Pearl Town, Shanghai

新湖明珠城地处上海市内环线以里，北起轻轨明珠线，东南至光复西路、苏州河。总用地28.11hm²，规划建设一个清新、自然、亲水傍绿的都市社区。

建筑单体以板式高层为主，由于苏州河有严格的退界高度限制，故建筑体形呈层层跌落状，整个小区错落有致，中央景观绿轴南北全长600余m，平均宽度60余m，它既是小区的"绿肺"，带来清新的空气与户户均好的景色，又是步行主轴，拥有树林、草坡、台地、水景、广场等一系列景观空间。

Located at the inner ring line of Shanghai and anchored at Mingzhu Line of light rail at north to Guang Fu Xi Lu and Suzhou River to the southeast side, the development covers a total site area of 28.11 hectares. Respectful of the planning requirements for the site by limiting the height of the buildings, the residence is mainly designed as the high-rise bars in the form of terraced structure. The central landscaping axis with a total length of approximately 600 meters from south to north and an average width of 60 meters is provided as the "green lung" to the site to allow for fresh air and from which residents could enjoy fantastic views. It also serves as the main pedestrian axis that is incorporated with trees, lawn, terrace, water features and plaza, etc.

建 设 单 位：上海新湖房地产开发有限公司
功 能 用 途：高层住宅
设计 / 竣工年份：2002 年 /2003 年
建 设 地 点：上海市普陀区东新路
总 建 筑 面 积 ：900500m²
总建筑高度 / 层数：100m/10～33 层
结 构 形 式：框剪结构

Client: Shanghai Xinhu Real Estate Development Co., Ltd.
Function: High-rise Residence
Designed/Completed: 2002/2003
Location: Dongxin Lu, Putuo District, Shanghai
Total Floor Area: 900,500m²
Total Height/Floor: 100m/10～33 Floors
Type of Structure: Framed Shear Wall Structure

方易西鉴枫景

Xijian Feng Jing, Hangzhou

方易西鉴枫景小区位于杭州市天目山路沿山河以北，炮台路以西。小区总用地面积为2.78hm²。

建筑总体布局合理，造型取自江南水乡建筑轻灵通透的神韵，以几何形的板式构件、大面积的玻璃和精致的金属栏杆、玻璃挡板等设计元素进行重组与穿插，强调立面和光线的虚实对比，直线的体块中不时地穿插入曲线的几何体，规整的结构系统和附加的构架极富韵律的组合突出了空间的变化。

The development covers a total site area of 2.78 hectares. The overall site plan is orderly defined. The form of the building evolves from the indigenous architecture in the Southern China. It is integrated and intersected by the geometric patterns, large block glass, delicate metal-clad balustrade and glass sidings to reinforce the visual contrast between the void and solid in the facades and from the light. Moreover the curved geometric patterns are integrated with the linear blocks and the design of formal structure and additional structural members emphasizes spatial diversity.

建 设 单 位：浙江大学科技园方易建设发展有限公司
功 能 用 途：住宅
设计 / 竣工年份：2003 年 /2005 年
建 设 地 点：浙江杭州天目山路
总 建 筑 面 积 ：100000m²
总建筑高度 / 层数：67m/20 层
结 构 形 式：框架、框剪结构

Client: Zhejiang University Scientific & Technological Park Fangyi Construction Development Co., Ltd.
Function: Housing
Designed/Completed: 2003/2005
Location: Mt. Tianmu Lu, Hangzhou, Zhejiang
Total Floor Area: 100,000m²
Total Height/Floor: 67m/ 20 Floors
Type of Structure: Framed & Framed Shear Wall Structure

湖州太湖人家

Taihu Lake Housing Development, Huzhou

太湖人家位于湖州市太湖以北，由太湖人家景观生态居住区及太湖人家度假村宾馆组成，总用地面积为14.6hm^2。居住区总建筑面积31817.1m^2，宾馆总建筑面积为14947m^2。设计通过一条贯穿整个小区曲线环路把小区各幢别墅包括宾馆组织起来，并以一条水系把整个小区贯穿起来，力图体现出湖州江南水乡的环境特色。

The 14.60-hectare site consists of a 31,817-square-meters landscaping ecological housing and a 14,947-square-meters Taihu Lake Resort Hotel. A curved approach road to the site is designed to integrate each villa and the hotel, while a river is master planned to meld the buildings into the landscaping environment that aims to address the vernacular aspects of Huzhou.

建设单位：湖州加业房产有限公司
功能用途：住宅、宾馆
设计/竣工年份：2002年/在建
建设地点：湖州市太湖以北
总建筑面积：46764m²
总建筑高度/层数：15m/3层
结构形式：框架结构

Client: Huzhou Jiaye Real Estate Development Co., Ltd.
Function: Residence, Hotel
Designed/Completed: 2002/under construction
Location: Taihu Lake (North), Huzhou
Total Floor Area: 46,764m²
Total Height/Floor: 15m/ 3 Floors
Type of Structure: Framed Structure

杭州湖畔花园北区

Lakeside Garden North Cluster, Hangzhou

建 设 单 位：杭州正大房产公司
功 能 用 途：住宅
设计 / 竣工年份：1998 年 /1999 年
建 设 地 点：杭州市文华路、竞舟路交叉口
总 建 筑 面 积：50000m^2；
结 构 形 式：框架、砖混结构

Client: Hangzhou Zhengda Real Estate Corp.
Function: Housing
Designed/Completed: 1998/1999
Location: Intersection of Wenhua Lu and Jingzhou Lu, Hangzhou
Total Floor Area: 50,000m^2
Type of Structure: Framed Structure, Brick & Concrete Structure

湖畔花园小区位于杭州蒋村商住区，总体设计以环境优美、户型合理为宗旨。住宅组团中心设一会馆，并设有180个车位的地下大型车库，车库上是6000m^{2}的绿化公园。

小区住宅户型较为多样，设计采用了大进深、跃层、退台等处理。底层架空部分为自行车库，部分作住户通高起居厅，并能由此进入南侧住宅庭园。而跃层及退台形成多种户型,有多个阳台和露台，使顶层户型更吸引人。

The principle concept for the design is to provide a landscaping environment and well-proportioned dwelling unit to the residents. A clubhouse is the centerpiece and a 6,000-square-meter garden is located above the basement car parking for 180 cars. Dwelling units of various sizes are provided with the maximum depth and some are designed as duplex-unit or "terraced" structures. Ground level is elevated to accommodate the bicycle parking and some are designed as the full height living room, from which residents can access the garden. Part of the residences are designed with duplex unit or step in that raises the variety of the residential types and offers numbers of balconies and terraces to make the dwelling units on top levels more inviting.

温州永嘉瓯北罗马城

Rome Town, Oubei, Yongjia County, Wenzhou

建设单位：温州新湖房地产开发有限公司
功能用途：商贸、住宅及配套会所
设计/竣工年份：2000年/2004年
建设地点：温州永嘉县瓯北镇阳光大道/罗浮大街
总建筑面积：112650m²
总建筑高度/层数：55.90m/18层
结构形式：框架剪力墙结构

Client: Wenzhou New Lake Real Estate Development Co., Ltd.
Function: Business, Housing, Supporting Club
Designed/Completed: 2000/2004
Location: Sunshine Boulevard/ Luofu Avenue, Oubei, Yongjia County, Wenzhou
Total Floor Area: 112,650m²
Total Height/Floor: 55.90m/ 18 Floors
Type of Structure: Framed Shear Wall Structure

温州永嘉瓯北罗马城紧邻瓯江北岸，地块南侧阳光大道为联系瓯江上各座大桥的主要交通干道，北面临瓯北镇中心，地块西侧罗浮大街为瓯北主要商业街的延伸段。基于对区域条件的分析，设计力求在住宅区内创造美好环境的同时，将业主要求设计的西侧近8000m²的商贸城有机地结合到整体设计中来，同时提供起伏错落的沿江建筑轮廓线，满足瓯江南岸的城市景观需要。总平面布局以一条南北贯通的步行商业街分隔商贸区和住宅区。与通常设计手法不同的是，地块南侧布置高层板式建筑，以提供最大的江景界面及体量标识，地块北侧则结合瓯北镇旧有的建筑尺度与中心景观带布置含有空中庭园的呈迭落状的多层建筑，从而保证了环境均好性。

Located adjacent to the northern bank of Oujiang River, the site is bounded to the Sunshine Boulevard which is the main road connecting several bridges over Oujiang River. The town center is at its north. Luofu Avenue at the west extends the town's main pedestrian street to the site. Through the analysis to the environment context around the site, the design for the development aims to provide a humane environment within the site and melds the 8,000-square-meter business town at the west into the site, while reflecting the silhouettes in the differentiated heights of the buildings facing the river. In the overall arrangement layout, a pedestrian street from the south to the north separates the business functions and the residential area. Different from the normal design, high-rise bars are displayed to the south allowing for the splendid view to the river and exhibiting its magnificent scale. To the north, integrating the old buildings and the central landscaping area provides stepped down multi-storey buildings with roof gardens.

嘉兴金穗太阳城

Jinsui Sun City, Jiaxing

嘉兴大树金穗太阳城位于嘉兴秀城新区骏力路东侧，基地为南宽北窄梯形，东侧为现有河道，基地面积10.8hm²，建筑内容包括别墅、多层住宅及公建。

总平面布局注重现有地域特点的利用，将现状的尽头河设计成一条回形内河，结合中心条形绿地设计。总平面路网采用中环式设计，别墅区以点式布置，尽端采用圆形的广场与步行道相结合的处理，使建筑物和周边的小环境更加开阔宜人。

Located in a trapezoid-shaped site with wide south and narrow north and bounded an existing river at the east, the 10.80hectare site includes the villa clusters, multi-story housing and public functions. Following the existing terrain within the site, the existing dead end river is master planned and transferred into a looped river. The circulation system is meticulously outlined as a middle ring route. Villas are clustered around the center. At the terminal a circular turn-around is integrated with the walkways to provide open and inviting amenities to the residents.

建 设 单 位：嘉兴大树房地产有限公司
功 能 用 途：住宅
设计 / 竣工年份：2003 年 /2004 年
建 设 地 点：嘉兴秀城区
总 建 筑 面 积 ：85788 m²
总建筑高度 / 层数：21m/3～6 层
结 构 形 式：砖混结构

Client: Jiaxing Dashu Real Estate Co., Ltd.
Function: Housing
Designed/Completed: 2003/2004
Location: Xiucheng District, Jiaxing
Total Floor Area: 85,788m²
Total Height/Floor: 21m/ 3～6 Floors
Type of Structure: Brick and Concrete Structure

杭州六合 · 天寓

Beyond City, Hangzhou

建 设 单 位：杭州六合房地产开发公司
功 能 用 途：住宅
设计 / 竣工年份：2003 年 / 在建
建 设 地 点：杭州滨江区
总 建 筑 面 积 ：250000m²
总建筑高度 / 层数：100m/33 层
结 构 形 式：框剪结构

Client: Hangzhou Liuhe Real Estate Development Corporation
Function: Housing
Designed/Completed: 2003/under construction
Location: Binjiang District, Hangzhou
Total Floor Area: 250,000m²
Total Height/Floor: 100m/ 33 Floors
Type of Structure: Framed Shear Wall Structure

本工程在总平面布置上打破了传统小区常用的建筑沿基地周边布置，集中绿化位于中央的封闭布局模式。设计采用开放的规划结构，即建筑物相对集中在基地中央，将绿化留在外围，把小区域的居住环境融入大区域的自然、人文环境中，同时也使得小区的绿化景观与周边的城市区域实现共享。

The overall site planning arrangement breaks the rule of enclosed type of configuration for the common housing development that the buildings are located around the roads with the garden in the center of the site. The planning principle is based on the concept of providing an open plan which the housing is clustered in the hub of the site and surrounded by the gardens and plants. This blends the living environment into the immediate context and shares the landscaping elements with the surrounding environment beyond the site.

杭州景江花苑

Jingjiang Garden, Hangzhou

建 设 单 位：浙江滨江建设有限公司
功 能 用 途：住宅、办公
设计／竣工年份：2003 年 /2005 年
建 设 地 点：杭州滨江 4 号区块
总 建 筑 面 积：120000m^2
总建筑高度／层数：99m/31 层
结 构 形 式：框剪结构

Client: Zhejiang Binjiang Construction Co., Ltd.
Function: Housing, Office
Designed/Completed: 2003/2005
Location: Lot No.4, Binjiang, Hangzhou
Total Floor Area: 120,000m^2
Total Height/Floor: 99m/ 31 Floors
Type of Structure: Framed Shear Wall Structure

景江花苑位于钱江三桥北岸一侧，堪称三桥的北岸"桥头堡"。项目由住宅公寓和单元式写字楼二幢"L"型高层建筑组成。地下设 2 层大型停车库。

高层住宅采用单元式一梯两户形式，户型设计中考虑到充分利用得天独厚的钱江景观，做到了"户户面江"、"户户朝南"最佳布置。高层写字楼采用串廊式平面，使每户写字间都拥有较好的朝向和良好的视觉景观。

Located at the northern bank of Qianjiang Third Bridge, this development is composed of apartment building and office building arranged in L shape with 2 basement levels for car parking. The architectural design for the apartment skyscraper allows for two dwelling units on each level as it takes advantage of magnificent views of Qiantang River that makes each unit to be oriented to the river and the south. The office tower is configured in gallery plan to offer the best orientation and perfect visual experience.

杭州同人·春江时代

Spring River Times • Tongren, Hangzhou

本项目位于杭州市滨江区长河镇，总用地 6.13hm²。总平面构图采用S型，营造出两个围合中央大绿洲，创造优良的生态环境。建筑集中布置，满足了城市小区的居住要求。设计上充分考虑了江景环境和生态概念，将整个地块标高由原来的6.5m抬高到9.80 m，同时临江设置联排两梯两户和二梯三户高层住宅和点式高层住宅，力求使更多的住宅具有良好的视野景观。

设计强调人车分流，汽车全部进入半地下车库和地下车库，充分保证住户生活的安全，中央大绿洲公园内配套设置不同年龄段的运动休闲项目，体现了以人为本的精神。

The development covers a total site area of 6.13 hectares and designed in S-shaped plan that creates two central lavishly planted courtyards and an ecological environment to the properties. The concept design for the project strives to capture the idea of Qiantang River and the ecological philosophy. The level of the site is raised to 9.80 meter from the former 6.50 meters. Detached town house is provided with two units and two stairs on one floor and high-rise building consists of two elevators and three units on one floor, together with the tower building so as to raise the number of the dwelling unit to enjoy the splendid view over to the river. The pedestrian circulation is clearly separated from the vehicle. Car access is restricted to a covered parking. The central garden is equipped with sport facilities for different ages that introduces a human-oriented design philosophy.

建 设 单 位：浙江同人置业有限公司
功 能 用 途：居住
设计 / 竣工年份：2004 年 / 在建
建 设 地 点：杭州滨江区
总 建 筑 面 积 ：229878m²
总建筑高度 / 层数：99.9m/33 层
结 构 形 式：框肢结构

Client: Zhejiang Tongren Real Estate Co., Ltd.
Function: Residence
Designed/Completed: 2004/under construction
Location: Binjiang District, Hangzhou
Total Floor Area: 229,878m²
Total Height/Floor: 99.90m/ 33 Floors
Type of Structure: Shear Wall Support on Frame Structure

义乌“银河湾”小区

Silver River Bay Housing Master Plan, Yiwu, Zhejiang

本项目位于义乌江畔，江滨路与新义中路交界处。基地总用地4.53hm²。设计从江南水乡风情出发，在满足功能要求的基础上致力于小区内水景的设置和绿化的营造。

居住区规划结构明晰，层次丰富，空间形态流畅。建筑轮廓呈流线形排列，与道路的走向一致，建筑与道路的有机结合，是本案的一大特色。设计利用基地内自然形成的小山丘，设置曲折路网，使之达到景观的变化，通过“隐、显、疏、密”的安排，达成江南园林“贵有层次，妙于曲折，在于深秀”的韵致。

Located near the bank of Yiwu River, the development approach to this 4.53-hectare site emphasizes the local vernacular aspects and incorporates the water and landscaping elements into the building while addressing functional requirements. All components within the site are thoroughly defined, diversified in the patterns and unified in spatial form. The design concept features the organic incorporation between the building and the roads. The architectural building is integrated into the roads which are meticulously outlines and deigned following the terrain of the site to achieve landscaping variety. The organization concept is to capture the essence of classic Chinese spatial planning principle which includes “concealment, disclosure, sparseness and thickness” that acknowledges the Chinese gardening ideology of “For mountains, it is crucial to have layers; For rivers it is the charm to be tortuous; For ridges it is superior to be graceful in the depth”.

建 设 单 位：浙江义乌银河房地产开发公司
功 能 用 途：住宅
设计 / 竣工年份：2005 年 / 在建
建 设 地 点：浙江义乌
总 建 筑 面 积 ：300392 m^2
总建筑高度 / 层数：39.0m/11 层
结 构 形 式：框剪结构、框架结构

Client: Yiwu Silver River Real Estate Development Corporation, Zhejiang
Function: Housing
Designed/Completed: 2005/under construction
Location: Yiwu, Zhejiang
Total Floor Area: 300,392m^2
Total Height/Floor: 39.00m/ 11 Floors
Type of Structure: Framed Shear Wall & Framed Structure

徐州汉泉山庄

Hanquan Villa Development, Xuzhou

四联别墅户型C-1效果图

四联别墅户型C-2效果图

项目位于徐州市铜山县，云龙湖以东，用地约81.6hm^2。基地依山傍水，黄河路把整个地块自然分为两部分，北区为独立别墅区，南区为联体别墅区。汉泉山庄作为徐州第一个高品质的大型别墅区，同时具有居住与休闲度假功能。

设计力图形成开放而又具有内聚力的空间景观格局，山景与水景渗透、延续，并形成由内而外、有疏有密、有高有低的空间层次，并最终与周围环境和谐统一。

The development is located at the east of Yunlong Lake of Tongshan County covering a total site of 81.60 hectares. Due to its splendid location surrounded by the mountain and the river, the site is divided into the independent villa clusters at the north and the detached villas at the south by Yellow River Road. As the first high-grade product of villa, the development features the functions for living and leisure. The architectural design strives to meld and extend the nature elements such as mountain and water into the building that achieves spatial diversity and formal richness and relates to the immediate context.

建设单位：徐州同创房地产开发有限公司
功能用途：住宅
设计／竣工年份：2003年／在建
建设地点：江苏徐州市铜山县
总建筑面积：401627m²
总建筑高度／层数：2～4层
结构形式：框架结构

Client: Xuzhou Tongchuang Real Estate Development Co., Ltd.
Function: Housing
Designed/Completed: 2003/under construction
Location: Tongshan County, Xuzhou, Jiangsu
Total Floor Area: 401,627m²
Total Height/Floor: 2～4 Floors
Type of Structure: Framed Structure

杭州中尚 · 橄榄树花园

Olive Garden • Zhongshang, Hangzhou

建 设 单 位：浙江中尚房地产开发有限公司
功 能 用 途：住宅
设计 / 竣工年份：2004 年 /2007
建 设 地 点：杭州市余杭区
总 建 筑 面 积：81341.5m²
总建筑高度 / 层数：36.2m/11 层
结 构 形 式：框架结构

Client: Hangzhou Zhongshang Real Estate Development Co., Ltd.
Function: Housing
Designed/Completed: 2004 / 2007
Location: Yuhang District, Hangzhou
Total Floor Area: 81,342m²
Total Height/Floor: 36.20m/ 11 Floors
Type of Structure: Framed Structure

项目位于杭州市余杭区，地块总用地面积4.8hm²，总建筑面积81341.5m²。在总体布局上以一条南北向弧形主车道贯穿用地，在南面和北面道路上设置主入口，中心绿地位于小区用地的中心。在东侧规划道路上设置一步行入口，并结合步行入口布置了一条贯穿中心绿地的东西向景观走廊。通过主路及景观走廊的划分，小区自然地形成了4个有机的组团。

The development covers a total site area of 4.8 hectares. The overall site planning arrangement provides a curved main driveway through the site and locates the main entrances at the south and north. A central planted garden is also provided. Pedestrian access is located at the eastern planned road from which a landscaping corridor extended from the east to the west passes through the central planted garden. Four clusters are organically configured and divided by the main approach and the landscaping corridor.

杭州人和家园

Renhe Garden, Hangzhou

建设单位：杭州人和房地产公司
功能用途：住宅
设计/竣工年份：2002年/2005年
建设地点：杭州余杭
总建筑面积：110000m²
总建筑高度/层数：50m/6～18层
结构形式：短肢剪力墙

Client: Hangzhou Renhe Real Estate Corporation
Function: Housing
Designed/Completed: 2002/ 2005
Location: Yuhang District, Hangzhou
Total Floor Area: 110,000m²
Total Height/Floor: 50m/ 6～18 Floors
Type of Structure: Framed Shear Wall Structure

杭州人和家园是集多种住宅类型于一体的舒适型居住小区。该区规划配套齐全,由四幢多层住宅、十幢10～11层小高层住宅、一幢17层高层住宅、一座幼儿园及其它商业等配套公建组成。

The concept design is based on the principle of providing fully equipped residential properties of various types. It includes four multi-storey residential buildings, ten 10 to11-storey residential buildings, one 17-storey high-rise residence, one kindergarten and other business functions.

山东聊城“城市主人”住宅区

“City’s Owner” Housing Development Liaocheng, Shandong

聊城“城市主人”住宅区是紧邻山东聊城东昌湖景区的大型亲水生态与节能社区，占地约 20hm^2。整个社区以生态节能为纲，东昌湖为主题，将别墅、联排别墅、花园洋房、多层、小高层住宅和商场、会所、临街商铺等配套设施加以组织，巧妙相融，空间恢宏，错落有致。

小区空间布局合理，层次分明，人车分流，景观设计体现造园韵味，精心设计的建筑造型与细节充分体现居住建筑在走向理性的同时，又注重对人性的全面关怀。

该项目被评为“齐鲁十大名盘”之一。

The development is bounded Dongchang Lake Resort and covers a total site area of 20 hectares. Respectful of the ecological and energy-saving concept and incorporating the immediate Dongchang Lake Resort into the development, the design challenge was to integrate the villa clusters, detached townhouse, low-storey residence, multi-storey residence, high-rise building, club house, shops at street level and the other supporting functions into the site. The site is organically master planned and well defined. Pedestrian circulation is clearly separated from the vehicles. Landscaping design strives to reflect the essence of Chinese gardening, while the well-proportioned and sophisticated facades and details exhibit the residential building towards the rationality and give the project a human scale. The development was awarded the honor of Well-known Project of Ten Residential Building Development in Qi Lu.

建 设 单 位：山东省南郊集团聊城三庆置业有限公司
功 能 用 途：排屋、别墅、中高层、多层住宅
设计 / 竣工年份：2005 年 / 在建
建 设 地 点：山东聊城东昌湖西侧
总 建 筑 面 积 ：225865m²
总建筑高度 / 层数：3～11 层
结 构 形 式：短肢剪力墙结构、砖混结构

Client: Shandong Southern Suburb Group Liaocheng TRI-CEL Real Estate Co., Ltd
Function: Townhouse, Villa, High-rise, Multi-storey Residence
Designed/Completed: 2005/under construction
Location: Dongchang Lake（West), Liaocheng, Shandong
Total Floor Area: 225,865m²
Total Height/Floor: 3～11 Floors
Type of Structure: Framed Shear Wall Structure, Concrete & Brick Structure

杭州江滨花园

Riverside Garden, Hangzhou

项目位于钱塘江、富春江、浦阳江三江汇合处。基地总用地 7.6hm²。小区设计以“环境、空间、文化、效益”为综合设计理念，创建了生态文化型景观广场。

结合基地环境因素，设计考虑项目自身特点，提炼出可供利用的五大界面：(1) 沿街商业休闲界面；(2) 水景商业与公共活动界面；(3) 湖面休闲界面；(4) 江景观赏界面；(5) 东方文化园景观界面，从而创造性地提出低密度、大围合、景观利用最大化作为本项目的规划主题。

Located at the intersection among Qiantang River, Fuchun Rive and Pujiang River, the proposed project covers a total site area of 7.6 hectares. The planning principle is based on the concept of integrating the environment, space, culture and the market value to create an ecological and cultural landscaping plaza.

Incorporating the immediate context and based on the characteristics of the development itself, the following aspects were taken into consideration:

- Business Leisure Facilities at Street Level.
- Business Facilities around the Water Features and the Public Space.
- Recreational & Entertainment Facilities on the Lake.
- View towards the River.
- View of Oriental Cultural Park beyond the Site.

The design approach thus proposes low density, great courtyard and maximum use of landscaping environment to the project.

建 设 单 位：绿都控股集团有限公司
功 能 用 途：商住综合楼
设计 / 竣工年份：2005 年 / 2009 年
建 设 地 点：杭州市萧山区
总 建 筑 面 积 ：246389m²
总建筑高度 / 层数：96.3m/30 层
结 构 形 式：框肢结构

Client: Ludu Holding Group Co., Ltd.
Function: Residence, Commercial
Designed/Completed: 2005/2009
Location: Xiaoshan, Hangzhou
Total Floor Area: 246,389m²
Total Height/Floor: 96.30m/ 30 Floors
Type of Structure: Support on Frame Structure

荆门紫薇苑

Ziwei Garden • Jinmen, Hubei

工程建于湖北荆门市，为一多层住宅项目。总平面布局中主要考虑到景观设计的合理性与共享性，争取绝大多数的住户都能享受到中心绿化景观带的景观。在户型设计上，主要考虑住户使用的便捷性、人性化，其中在沿景观带的边套设计了复式户型。立面及造型设计上，主要以阳台、空调机位的合理自由组合来体现立面的丰富多样性。

The overall site planning arrangement for this multi-storey dwelling development emphasizes the rationality and integration of the landscaping design. It allows for the most dwelling units enjoying the magnificent view towards the central green area. The concept design for the type of the dwelling unit addresses the requirement of accessibility and human-oriented philosophy. Balconies are provided to be incorporated with the location for the air-handling units to reinforce the variety and diversity of the facades and form of the building.

建 设 单 位：荆门紫薇房地产有限公司
功 能 用 途：住宅
设计 / 竣工年份：2004 年 / 2006 年
建 设 地 点：湖北荆门
总 建 筑 面 积：40000m²
总建筑高度 / 层数：19m
结 构 形 式：砖混结构

Client: Hubei Jinmen Ziwei Real Estate Corporation
Function: Housing
Designed/Completed: 2004 / 2006
Location: Jinmen, Hubei
Total Floor Area: 40,000m²
Total Height/Floor: 19m
Type of Structure: Brick & Concrete Structure

铜陵世纪曙光苑

Aurora Century Garden, Tongling, Anhui

建 筑 单 位：铜陵市营佳房地产有限公司
功 能 用 途：住宅
设计 / 竣工年份：2004 年 / ~
建 设 地 点：安徽省铜陵市
总 建 筑 面 积：40578m²
总建筑高度 / 层数：18 层
结 构 形 式：框剪结构

Client: Tongling Yingjia Real Estate Co. Ltd.
Function: Housing
Designed/Completed: 2004/~
Location: Tongling, Anhui
Total Floor Area: 40,578m²
Total Height/Floor: 18 Floors
Type of Structure: Shear Wall Structure

项目采用围合式布局，西侧布置三幢点式高层，北面一幢点式高层和一幢中高层沿街排布，南面安排一幢多层，内部围合成一个幽静的庭院，与西侧的义安公园相呼应，使整个小区的内外都有最好的景观。交通采用人车分流的形式，机动车从小区入口直接进入地下车库，而小区内部则实行完全的步行系统，避免机动车对住户的干扰。

The quadrangle composition consists of 3 high-rise towers at the west, one high-rise tower at the north, one medium-storey building at the street level and one multi-storey at the south. The central landscaping garden that corresponding to the Yi'an Park at the west provides welcoming amenities to the residences. Vehicular traffic to the site is separated from the pedestrian circulation and has direct access to a covered parking to maximize the open pedestrian space above ground.

杭州青春都市公寓

Young Urban Condominium, Hangzhou

建 设 单 位：杭州商宇房产开发有限公司
功 能 用 途：住宅
设计/竣工年份：2004年/2008年
建 设 地 点：杭州南肖埠原杭州钢窗厂地块
总 建 筑 面 积：27072m²
总建筑高度/层数：39m/10层
结 构 形 式：框剪结构

Client: Hangzhou Shangyu Real Estate Development Co., Ltd.
Function: Housing
Designed/Completed: 2004 / 2008
Location: Nanxiaobu, Hangzhou
Total Floor Area: 27,072m²
Total Height/Floor: 39m/ 10 Floors
Type of Structure: Framed Shear Wall Structure

本工程由3幢小高层和1幢多层组成，4幢住宅围合成一个集中绿地，其中的2幢底层架空，使绿化得以渗透，景观能够延续。

The development contains three medium-storey buildings and one multi-storey building that encloses a central garden. To meld the green elements into the dwelling buildings and extend the landscaping features, the ground floor of two buildings is elevated.

体育建筑
Sports Building

杭州黄龙体育中心

Dragon Sports Center, Hangzhou

主体育场平面形状外圆内椭，内场设一400m跑道标准田径场及标准足球场，看台东西高、南北低。看台上部的罩棚为复合结构，由南北双塔用斜拉索拉住东西两片罩棚（空间网壳）。体育中心建筑造型新颖别致，简洁流畅，气势雄伟，体现了现代体育建筑的特色。在平面功能上，将观众流线与其它流线用高架休息平台进行了分离，较好地解决了流线相互干扰问题。设计充分利用看台下部空间，作为办公、娱乐、酒店等，给体育场带来了良好的经济效益。

主体育馆位于主体育场的东北侧，建筑造型通过四片弯月形屋面的巧妙组合，使自身既富有强烈的个性，充满现代感和运动感，又使其与主体育场比较协调。在平面功能上，将训练馆与比赛馆合二为一，平面为椭圆形，比赛馆内场尺寸以冰球场为基础，设观众座席8000座（其中1500座为活动座席）。

The 60,000-seat main stadium accommodates a standard athletic field with a 400-meter-long track and a typical football field. A composite structure of canopy is provided at the west and east sides and supported by the stayed cables anchored to the dual towers at the north and south ends. The west-east stand is located at high level and the north-south stand at low level. The appearance of the stadium is simple and innovative conveying a sense of power and times that captures the nature of the sports building. In the plan layout design, the pedestrian circulation is meticulously outlined and separated from the other circulation by the raised terraces. Under the stands the office, entertainment and hotel facilities are provided.

The main gymnasium is located at the northeast side of the main stadium and consists four crescent-shaped roof structures to enforce its identity that conveys a sense of fashion and movement and echoes to the main stadium. In the plan layout design the training halls and venues are located under one roof and designed in the shape of ellipse. The inner dimensions of the venues are determined based on the scale of the ice hockey field and provided with 8000 seats including 1500 moveable seats.

建 设 单 位：黄龙体育中心建设办公室
功 能 用 途：体育比赛
设计 / 竣工年份：体育场：1996 年 /2000 年 ， 体育馆：2000 年 /2003 年
建 设 地 点：杭州市黄龙洞风景区以北
总 建 筑 面 积 ：102000m^2 · 体育馆：25000m^2
总建筑高度 / 层数：体育场：85m /22 层， 体育馆：39m /2 层
结 构 形 式：框架、空间网壳结构

Client: Task Force of Hangzhou Dragon Sports Center
Function: Sports
Designed/Completed: Main Stadium1996/2000,Gymnasium 2000/2003
Location: Dragon Cave Scenic Resort（North）, Hangzhou
Total Floor Area: Main Stadium 102,000m^2 ,Gymnasium 25,000m^2
Total Height/Floor: Main Stadium 85m/22 Floors, Gymnasium39m/ 2 Floors
Type of Structure: Framed & Spatial Shell Structure

义乌市会展体育中心体育馆

Convention, Exhibition & Sports Center Gymnasium, Yiwu

义乌市会展体育中心位于义乌市宾王大桥以东，设有主体育场、体育馆、游泳馆、会展中心以及附属用房等。体育馆位于中心的东侧，设有6000个观众座席。建筑造型设计采用了两个不同大小的半圆形的体量组合形式。屋顶为与平面对应的两组壳体，均为规则的几何体块。屋面采用新型的半透明的聚碳酸酯板，使在白天室内的采光及晚上的夜景俱佳，建筑造型简洁有力，富有现代感。

The development is composed of main stadium, gymnasium, natatorium, convention & exhibition center and supporting functions. Located at the east of the site and provided with 6000 seating capacity, the gymnasium is designed with two composite semi-circular volumes of different sizes. The roof employs two pairs of shell to define its geometry that echoes to its plan. The materials to be used on the roof are translucent PTFE panels to achieve the effective lighting in the daytime and a splendid view at night.

建 设 单 位：义乌市体育中心指挥部
功 能 用 途：体育
设计 / 竣工年份：2000 年 /2003 年
建 设 地 点：浙江义乌市宾王大桥以东
总 建 筑 面 积 ：38200m²
总建筑高度 / 层数：36.5m/3 层
结 构 形 式：框架（顶部为钢结构）

Client: Yiwu Sports Center Headquarter
Function: Sports
Designed/Completed: 2000/2003
Location: Binwang Bridge (East), Yiwu
Total Floor Area: 38,200m²
Total Height/Floor: 36.50m/3 Floors
Type of Structure: Framed Structure, Steel Structure Roof

中国棋院杭州分院

Chinese Chess Institute Hangzhou Branch, Hangzhou

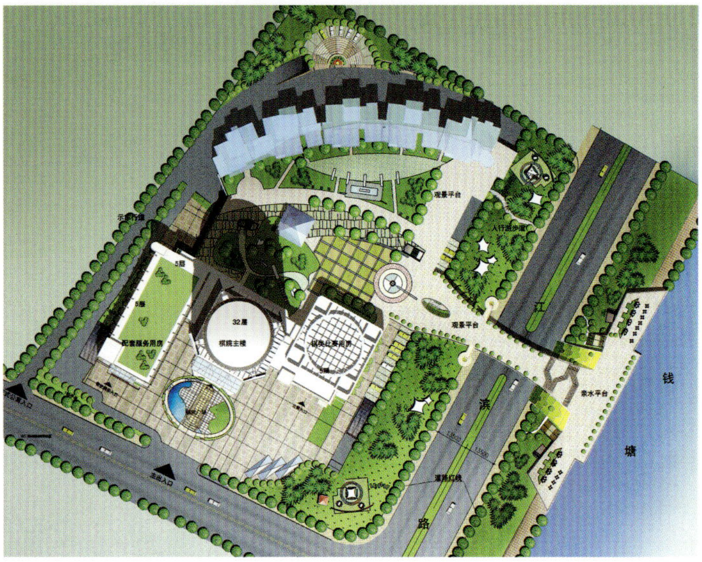

建 设 单 位：杭州棋院
功 能 用 途：棋、牌类比赛训练学校、宾馆
设计 / 竣工年份：2002 年 /2005 年
建 设 地 点：杭州钱江新城
总 建 筑 面 积 ：56719m^2
总建筑高度 / 层数：145.3m/ 34 层
结 构 形 式：框筒结构

Client: Hangzhou Chess Institute
Function: Training School for Chess & Card Competition, Guest House
Designed/Completed: 2002/2005
Location: Qianjiang New City, Hangzhou
Total Floor Area: 56,719m^2
Total Height/Floor: 145.30m/34 Floors
Type of Structure: Framed Cored-wall Structure

本工程位于杭州钱江新城，占地面积 1.64hm^2。中国棋院杭州分院是拥有比赛中心、围棋学校、围棋培训中心、生活配套用房、全民健身附属设施等多功能综合性的棋院。地块东侧布置 5 层群众活动中心和比赛中心，既能单独对外开放使用，又方便与主楼联系。地块中心设置 34 层主楼，内设围棋学校、三棋训练用房、运动员公寓及贵宾客房等，地块西侧设 4 层餐饮中心。

The multi-used development covers a site area of 1.64 hectares and houses the game center, chess school, chess training center, supporting functions for living and auxiliary sports facilities for public. Beyond the east side of the site includes the 5-storey pubic athletic center and competition center which could be open to public use and also easily connected to the 34-storey main tower that is strategically located in the center and contains the chess school, training premises, apartment for the athletes and VIP guest rooms. A 4-storey restaurant is served at the west.

宁波市鄞州体育馆

Yinzhou Gymnasium, Ningbo

建 设 单 位：宁波市鄞州中心区管委会
功 能 用 途：体育馆
设计 / 竣工年份：2002 年 /2004 年
建 设 地 点：宁波市鄞州高教园区
总 建 筑 面 积：27264m²
总建筑高度 / 层数：22m/3 层
结 构 形 式：框架 + 钢屋顶结构

Client: Ningbo Yinzhou Center Administrative Committee
Function: Sports
Designed/Completed: 2002 / 2004
Location: High Education Park, Yinzhou, Ningbo
Total Floor Area:27,264m²
Total Height/Floor: 22m/3 Floors
Type of Structure: Framed Structure, Steel Structural Roof

工程位于宁波市鄞州区高教园区内，规划用地面积为 8.9hm²。用地形状为平行四边形，其东、北两侧有河水围绕。本工程由 4000 座的圆形体育馆及三角形的训练馆两个部分组成，造型简洁大方，现代气息浓郁。体育馆为多功能型，内场尺寸为 50m × 30m，观众席平面为椭圆形，以使更多的观众席拥有良好的视觉质量。

The multi-functional development is located in an 8.90-hectare parallelogram-shaped site surrounded by the rivers at its east and north side. It contains a circular multi-used 4000-seat gymnasium and a triangular-shaped training hall, the former is designed with an inner dimension of 50 meters long and 30 meters wide. The crescent-shaped stand ensures that everyone has an unobstructed view from each. The simple form of the building gives itself a modern image.

杭州游泳健身中心

Hangzhou Aquatic Athletic Center, Hangzhou

杭州游泳健身中心位于杭州市中心武林广场区域内，北临京杭大运河，为杭州古城墙的北段所在。整个项目由一座具备举办全国游泳竞赛要求的游泳馆和一幢体育健身活动楼组成，波浪型的白色屋面反映游泳运动特色，缓缓上升的建筑造型体现体育事业的稳步发展，与传统建筑神似的双曲面双坡顶的构造又隐寓着运河沿岸1500年的繁华。屋面结构在全国首度采用轻盈的双层骨架膜结构，通过室内自然气流的组织和引导湿气的排放，有效控制馆内结露现象的产生，具有明显的节能效果。1999年全国游泳锦标赛暨奥运选拔赛在此成功举办，该项目曾获2000年浙江省优秀设计一等奖和全国工程“鲁班奖”。

Located in the precinct of downtown Wulin Square and surrounded by the buildings of various kinds, the development is bounded Beijing-Hangzhou Grand Canal at north. Due to its strategical location, the architecture of the complex should pay homage to the magnificent cultural heritage around it. This multi-used complex accommodates a natatorium to be designed hosting national swimming games and an athletic sport building. The soft undulating curves of the natatorium are complemented to the slowly upward form, which is reminiscent of the swimming sports and the stable development of sports culture. The roof structure completed with dual curved planes and double pitched-roof that is similar to the traditional architecture is a metaphor of prosperity around the Grand Canal over 1500 years. Roof level is designed with light double-layered trussed membrane that exhausts the moisture through the indoor natural air circulation system and controls the dew occurrence. It is confirmed to be an effective energy conservation design. In 1999 it hosted the National Swimming Championships and Olympic Selection Games. The development was awarded the honor of First Prize of Zhejiang Excellent Design 2000 and Luban Prize China 2000.

建设单位：杭州市体育运动委员会
功能用途：游泳馆及体育健身活动
设计／竣工年份：1998 年／1999 年
建设地点：杭州市环城北路　中山北路
总建筑面积：32000m²
总建筑高度／层数：46.6m／地上 7 层，地下 2 层
结构型式：框架结构、屋面钢结构骨架膜

Client: Hangzhou Athletic Sports Committee
Function: Natatorium, Athletic Sports Facilities
Designed/Completed: 1998/1999
Location: Ring Road（North）/ Zhongshan Lu（North), Hangzhou
Total Floor Area: 32,000m²
Total Height/Floor: 46.60m / 7 Floors（aboveground), 2 Floor （below ground）
Type of Structure: RC Framed Structure, Roof Steel Structural & Trussed Membrane

台州市椒江文化体育发展中心

Jiaojiang Cultural & Sports Development Center, Taizhou

椒江文化体育发展中心系一个现代的多功能健身、比赛、娱乐场所，内设3000座的体育馆、训练馆、游泳馆、文体培训中心、娱乐用房等。建筑总用地面积3.48hm^2。项目在整体布局上以3000座的体育馆为中心，训练馆、游泳馆、文体培训大楼等围绕其成扇形布置，建筑错落有序并在体育馆之间形成一条流动的景观平台，也是各场馆之间联系的桥梁。地块的西南侧设置了网球场、篮球场等，同时作为西侧居住区的健身场所。在立面造型上采用了现代简洁的设计手法，着重表现体育建筑的力度感，并对体育馆、游泳馆等大空间采用了贝壳状的屋面造型，以体现当地的滨海城市特性。

This state-of-the-art multi-used facility covers a total site area of 3.48 hectares and houses a 3000-seat gymnasium, athletic training hall, natatorium, cultural sports training center and entertainment functions. In the overall plan arrangement the training hall, natatorium and the cultural sports training center are configured around the central gymnasium in a fan-shaped plan and connected to the gymnasium as a fluid landscaping terrace and bridge between them. Tennis court and basketball playground are served at the southwest side as the physical exercise place for the residence at the west. The facades of the building are designed to exhibit its modern simplicity and strive to embody the power of the sport architecture. Shell-shaped roof level is provided to the gymnasium and natatorium acknowledging the indigenous seafront aspects of the building.

建设单位：椒江文体中心工程建设指挥部
功能用途：体育比赛、文化娱乐
设计/竣工年份：2004年/2007年
建设地点：台州市椒江区
总建筑面积：42530m²
总建筑高度/层数：60m/16层
结构形式：框架结构

Client: Taizhou Jiaojiang Cultural & Sports Development Center Headquarter
Function: Sports Game, Cultural Entertainment
Designed/Completed: 2004 / 2007
Location: Jiaojiang District, Taizhou
Total Floor Area:42,530m²
Total Height/Floor: 60m/16 Floors
Type of Structure: Framed Structure

舟山市海洋体育文化公园

Zhoushan Sea Sports & Cultural Park, Zhoushan

建 设 单 位：浙江省舟山市城乡建设委员会
功 能 用 途：体育
设计 / 竣工年份：2004 年 /2007 年
建 设 地 点：浙江省舟山市
总 建 筑 面 积 ：95500m^2
总建筑高度 / 层数：33.6m/3 层
结 构 形 式：框架、框剪、大跨度空间结构

Client: Zhejiang Zhoushan Urban & Rural Construction Committee
Function: Sports
Designed/Completed: 2004/2007
Location: Zhoushan, Zhejiang
Total Floor Area:95,500m^2
Total Height/Floor: 33.60m/3 Floors
Type of Structure: Framed Structure，Framed Shear Wall Structure，Large-spanned Spatial Structure

工程位于舟山海洋文化名城，周边为市民中心及海洋文化公园。项目包括一个3万人中型体育场，一个3000人中型体育馆，以及游泳馆、网球馆和综合训练馆等相关体育运动场馆。为体现舟山海岛文化特征以及城市标志性建筑的特色，体育场采用简洁的造型设计，就如同阳光下海边闪烁的贝壳，而游泳馆、网球馆、综合训练馆则以体育馆为中心，以半圆形放射状围绕其布置，既像海滩上的海螺，又像飞旋海面上的浪花，以流畅透明的玻璃表达海水的晶莹，整个建筑群洋溢着向上的动感和热情。

Located in Zhoushan Sea Cultural Town and surrounded by the Civic Center and Sea Cultural Park, the multi-functional development is composed of a 30,000-capacity stadium, a 3,000-capacity gymnasium, natatorium, tennis hall, comprehensive training halls and other related athletic sport halls. In order to acknowledge the island cultural aspects of Zhoushan and create a signature building to the city, the simple form of the stadium symbolically relates to the twinkling shell in the sunshine beach, while the gymnasium is surrounded by the natatorium, tennis hall and training halls in a fan-shaped plan reminiscent of the conch or the spray of the waves. The transparent glass materials are used as a metaphor of sparkling water. The design of the architecture conveys a sense of lightness, enthusiasm and flight.

嘉兴市残疾人奥林匹克中心

Jiaxing Paralympic Games Center, Zhejiang

建 设 单 位：嘉兴市残疾人联合会
功 能 用 途：体训、宾馆
设计 / 竣工年份：2005 年 /2007 年
建 设 地 点：嘉兴市
总 建 筑 面 积：30000m²
总建筑高度 / 层数：78m/19 层
结 构 形 式：框架结构

Client: Union for the HandicappedJiaxing
Function: Physical Exercise, Hotel
Designed/Completed: 2005/2007
Location: Jiaxing, Zhejiang
Total Floor Area: 30,000m²
Total Height/Floor: 78m/ 19 Floors
Type of Structure: Framed Structure

嘉兴市残奥中心为多功能的比赛、训练、公共活动场所，含体训馆、辅助训练馆、（宾馆客房 200 间）等。设计注意残疾人的特殊要求，建筑集中布置，建筑群体通过体量与曲直线条的对比与组合，体现“拼搏、进取”的体育精神与建筑内涵。

This multi-functional center combines the venue, physical training and public activity under one roof. It includes the physical training halls, supporting training halls and a 200-room hotel, etc. The architectural design of the building focuses on the specific requirement of the handicapped. It is designed in a centralized configuration which is complemented with its massing and the contrast between curve and straight lines that embodies the combatant and active spirit of sport.

台州体育中心规划及主体育场

Taizhou Sports Center Master Plan & Main Stadium, Taizhou

台州市体育中心位于台州市中心区。其主要建筑内容为一场三馆及相应配套设施，即40000座的主体育场、4500座的体育馆及训练馆、3000座的游泳馆和2500座的网球馆等。主体育场位于用地的中心，是体育中心的主体。考虑台州市为港口城市，且体育中心靠近东海，体育场设计采用了"扬帆起航"的设计立意，采用轻盈、悬浮的看台罩棚，高耸的桅杆，紧绷的悬索，流动波浪般的看台等建筑造型，使体育场看起来就象是一艘扬起风帆，就要起航的巨舰。在体育场的东北角方向设置了1/4圆弧的大平台，与东北侧的入口广场及45°轴线方向的喷水池、绿地、广场相呼应，更衬托了体育场的宏大气势。

Located at the downtown of Taizhou, the sports center houses a 4,000-seat main stadium in the center, a 4,500-seat gymnasium and training hall, a 3,000-seat natatorium and a 2,500-seat tennis court, etc. Considering that Taizhou is a coastal city and the site is near the East Sea, the concept design of the stadium uses light and floating canopy over the stand, up straight masts, stiffened ropes and undulating curve of stand to convey a sense of sailing cruiser. A fan-shaped terrace of monumental scale is located at the northeastern corner corresponding to the entrance concourse at the northeastern side, the water fountain, the planted area and the square that enhances the magnificent and grandiose presence of the stadium.

建设单位：台州市体育中心建设指挥部
功能用途：体育
设计/竣工年份：1999年/2003年
建设地点：浙江省台州市
总建筑面积：49090m²
总建筑高度/层数：70m/3层
结构形式：框剪结构（屋顶钢结构、斜拉索）

Client: Taizhou Sports Center Construction Headquarter
Function: Sports
Designed/Completed: 1999/2003
Location: Taizhou, Zhejiang
Total Floor Area: 49,090m²
Total Height/Floor: 70m/ 3 Floors
Type of Structure: Framed Shear Wall Structure, Roof Structural Steel, Stayed-cable

临海体育中心体育馆

Linhai Sports Center Gymnasium, Zhejiang

建设单位：临海体育中心建设指挥部
功能用途：体育
设计/竣工年份：2000年/2003年
建设地点：临海市新城中心南湖边
总建筑面积：31365 m²
总建筑高度/层数：37m /3层
结构形式：框架结构、钢结构

Client: Linhai Sports Center Construction Headquarter
Function: Sports
Designed/Completed: 2000/2003
Location: South Lake, New City Center, Linhai, Zhejiang
Total Floor Area: 31,365m²
Total Height/Floor: 37m/ 3 Floors
Type of Structure: Framed & Steel Structure

临海市体育中心位于临海市新城市中心南湖边，区内设有一场二馆（4000座主体育场、6000座体育馆、4000座游泳馆），总用地面积24.53hm²，体育馆为一期工程，总建筑面积31365m²，建筑外部造型力求体现滨水建筑与体育建筑的特点，烘托出动感和“风帆”的主题，并采用新型的材料体现艺术与技术的完美结合，使其成为临海市的标志性建筑。

The sports center is composed of a 4,000-seat main stadium, a 6,000-seat gymnasium and a 4,000-seat natatorium. It covers a total site area of 24.53hectares. The appearance of the first phased gymnasium features the shape of the Sail in the Wind that conveys a sense of dynamic movement and portrays the image of waterfront development and the sports architectures. Building architecture also strives to reflect new technology in building systems and materials. Upon completion it serves as an icon building in the city.

安吉体育文化中心

Anji Sports & Cultural Center, Zhejiang

建设单位：安吉生态园区有限公司
功能用途：体育、会展
设计/竣工年份：2003年/2004年
建设地点：安吉市灵峰路
总建筑面积：27000 m²
总建筑高度/层数：32m/1层
结构形式：大跨度空间结构

Client: Anji Ecological Park Co., Ltd.
Function: Sport, Convention, Exhibition
Designed/Completed: 2003/2004
Location: Lingfeng Lu, Anji
Total Floor Area: 27,000m²
Total Height/Floor: 32m/ 1 Floor
Type of Structure: Large-spanned Spatial Structure

安吉体育文化中心位于灵峰路与天荒坪南路交叉口三角地块。体育馆规模为3000座，平面布局为椭圆形。文化馆的主体是大型展览中心，设于体育馆东侧，并通过办公区域、中庭与体育文化中心主体相连。

在建筑造型上采用椭圆平面，两层叠壳，更显舒展轻盈。屋顶有力地向外伸展，蕴含着热情与力量，两个球形突破上层屋顶，犹如两颗呼之欲出的明珠正日益绽放出光辉。

The development consists of two main functions. The ellipse-shaped stadium holds 3000 spectators. Cultural Hall is dominated by the exhibition center at the east of the stadium and connected to the main structure of the center through the office complex and atrium. The bold form of the roofing extends beyond it to express the enthusiasm and strength as the game itself. Two sphere-shaped roof structures are protruded towards the sky as if two bright pearls will be jumped out.

杭州之江饭店

宁波东港大酒店

常州大酒店

萧山国际大酒店

浙静中日友好会馆

浙江玉环大酒店

绍兴饭店

普陀山息来小庄

杭州临平宾馆

杭州萧山国际机场基地综合楼

山东莱芜市政府

中国水稻研究所

上海浦东机场办公楼

湖州电力调度中心

秦山核电公司检修培训文档中心

杭州耀江广厦

上海银行培训中心

杭州华浙广场

宁波外经贸大楼

杭州日报大楼

上海瑞安广场

上海兰生大厦

杭州富春大厦

玉环县建设环保大楼

上海斯波特酒店

浙江证券公司

浙江玉环县科技文化中心

杭州汽车东站

中国丝绸博物馆

浙江省老年活动中心

余姚河姆渡遗址博物馆

浙江音乐厅

杭州市体育馆（原浙江省体育馆）

慈溪体育场

杭州高级中学

杭州师范学院

上海瑞南新苑

杭州新世纪高层住宅

上海鸿发苑

杭州国际花园

杭州西湖花苑

杭州国信花园

上海老西门商业住宅区

杭州罗马公寓

广东中山市人民医院外科楼

常州第一人民医院病房楼

东阳巍山医院

慈溪人民医院

上海卢湾区中心医院

浙江邮电医院

苏州医学院第一附属医院外科病房楼

广东佛山市中医院

上海东方医院

浙江关怀医院

《浙江省建筑设计研究院作品选》编委会

主　　任：景政治
副 主 任：施祖元
学术顾问：益德清　唐葆亨　方子晋
执行主编：蔡金福
编　　委：曹跃进　李志飙　许世文
　　　　　杨学林　蒋　纹　蔡　彤
翻　　译：毛　冰
校　　对：兰曼利
摄　　影：沈欢顺　蔡金福
责任编辑：郭洪兰
封面设计：方舟正佳